Synthesis Lectures on Sustainable Development

Series Editor

Thomas Siller, Civil and Environmental Engineering, Colorado State University, Fort Collins, USA

This series publishes short books related to sustainable development practices relevant to engineers, technologists, managers, educators, and policy makers. The books are organized around the United Nations Sustainable Development Goals for 2015–2030. Design for sustainability along with the economics of sustainable development will be the common themes for each book. Topics to be covered will span all the major engineering disciplines along with the natural and environmental sciences that contribute to an understanding of sustainable development. The goal of this series is to make theory and research accessible to practitioners working on sustainable development efforts.

Vijay Prakash Gupta · A. K. Haghi ·
Anuradha Yadav

Green IoT and AI for Sustainable Development of Smart Cities

Springer

Vijay Prakash Gupta
Institute of Business Management
GLA University
Mathura, Uttar Pradesh, India

A. K. Haghi
Edinburgh, UK

Anuradha Yadav
DPG Degree College
Gurugram, Haryana, India

ISSN 2637-7675 ISSN 2637-7691 (electronic)
Synthesis Lectures on Sustainable Development
ISBN 978-3-031-98742-7 ISBN 978-3-031-98743-4 (eBook)
https://doi.org/10.1007/978-3-031-98743-4

This Springer imprint is published by the registered company Springer Nature Switzerland AG
The registered company address is: Gewerbestrasse 11, 6330 Cham, Switzerland

Preface

In a time where swift urbanization is redefining how we live, Green Internet of Things (Green IoT) and Artificial Intelligence (AI) are leading the way to create sustainable, efficient, and livable smart cities. They are not merely instruments—they are the foundation of a new era in urban construction, where innovation and ecology join hands.

Green IoT and AI are leading the transformation of contemporary cities by providing more intelligent solutions for energy consumption, transport, waste, governance, and citizen engagement. They enable the creation of cities that are not just intelligent and efficient, but also green and responsive to the needs of their citizens.

These "green" technologies are the pillars of smart cities that are founded on the basis of comfort, convenience, connectivity, and sustainability. Globally, public leaders, corporate entrepreneurs, and IT experts are harnessing advanced technologies—such as AI-powered analytics, energy-saving IoT networks, and digital infrastructure—to solve essential urban problems. Together, their efforts are making the quality of life better for citizens and shrinking the ecological footprint of cities.

This book offers an in-depth discussion of recent developments in Green IoT and AI technologies and their application to the Sustainable development of smart cities. It discusses the engineering complexities and real-world challenges involved in implementing these technologies at scale. Through case studies, new research, and future-oriented perspectives, the book provides useful insights to researchers, policymakers, urban planners, and technology leaders. Lastly, this book acts as an invaluable reference for seeing how the harmonious integration of Green IoT and AI can bring resilient, future-proof cities—marking a future of more intelligent, green, and accessible cities.

Mathura, India — Vijay Prakash Gupta (Lead Author)

Edinburgh, UK — A. K. Haghi

Gurugram, India — Dr. Anuradha Yadav

Declaration

On behalf of all the authors—Dr. Vijay Prakash Gupta (Lead Author), Dr. K. Haghi, and Dr. Anuradha Yadav—and with their full consent, we declare the following:

This declaration is concerning the use of artificial intelligence (AI) in chapter manuscripts which is as follows. While we do not encourage or permit the use of AI-generated content, AI-assisted tools may be used for language refinement and structuring. To ensure transparency, on the behalf of authors this confirms that no artificial intelligence (AI)-generated content or AI-assisted writing tools were used in the preparation of this chapter (including summaries and the abstract). All text, analysis, and conclusions are the result of the author's original work. The author acknowledges the use of AI-based tools for minor language editing or grammatical correction. However, all intellectual contributions, analyses, and conclusions presented in this chapter are entirely the author's own.

Signed on behalf of all authors,

Dr. Vijay Prakash Gupta (Lead Author)
May 2025

Contents

1 Components and Requirements for Digital Infrastructure 1
1.1 The Role of Digital Infrastructure in Smart Cities 3
1.1.1 Big Data and Predictive Analytics for Smart Cities 4
1.1.2 Green Internet of Things (G-IoT) and Sustainability 5
1.1.3 Artificial Intelligence (AI) and Machine Learning in Smart Cities 5
1.1.4 The Future of Digital Infrastructure for Smart Cities 6
1.2 Big Data for a Sustainable Smart City 7
1.2.1 The Role of Big Data in Smart Cities 7
1.2.2 The Importance of Predictive Analytics 8
1.2.3 The Role of Big Data in Smart Cities 8
1.2.4 Big Data for Environmental Sustainability 9
1.2.5 Big Data for Urban Mobility and Transportation 11
1.3 G-IoT for Sustainable Smart Cities and Society 12
1.3.1 Defining Green Internet of Things (G-IoT) 12
1.3.2 Practical Applications Through Case Studies 14
1.3.3 Interdisciplinary Ideas for Smart City Development 15
1.3.4 Integrating Technology and Sustainability Concepts 15
1.3.5 The Role of G-IoT in Smart City Sustainability 15
1.3.6 Case Studies: G-IoT in Action 16
1.4 Innovative Approaches in Environmental Monitoring 16
1.4.1 Real-Time Environmental Monitoring 16
1.4.2 The Role of IoT in Environmental Monitoring 17
1.4.3 Remote Sensing and Satellite Imagery 17
1.4.4 Artificial Intelligence and Big Data Analytics 17
1.4.5 Citizen Science and Crowdsourcing 18
1.4.6 Drones and Unmanned Aerial Vehicles (UAVs) 18
1.4.7 Blockchain for Environmental Data Integrity 19

1.4.8 Challenges and Future Directions 19
1.4.9 The Use of AI in Environmental Monitoring 19
1.5 Green Infrastructure and Urban Adaptation 20
1.5.1 Green Infrastructure for Sustainable Cities 20
1.5.2 The Role of Green Infrastructure in Urban Adaptation 21
1.5.3 Benefits of Green Infrastructure 21
1.5.4 Innovative Strategies for Green Infrastructure Implementation 22
1.5.5 Case Studies of Green Infrastructure Success 22
1.5.6 Challenges in Implementing Green Infrastructure 23
1.5.7 The Future of Green Infrastructure 23
1.5.8 Urban Adaptation to Climate Change 24
1.5.9 Case Study: The "Sponge City" Concept 24
1.6 Smart Cities Management and Innovations for Sustainable Development 24
1.6.1 AI and Machine Learning in Smart Cities Management 24
1.6.2 Innovations in Smart City Governance 25
1.6.3 Integrating Technology and Sustainability 25
1.6.4 The Pillars of Smart Cities 26
1.6.5 Innovations in Smart City Management 26
1.6.6 Sustainable Development via Smart Cities 27
1.6.7 Case Studies in Smart Cities Management 27
1.6.8 Challenges in Smart Cities Management 28
1.6.9 Future Directions in Smart Cities Management 28
References 29

2 Green Internet of Things (G-IoT) for Sustainable Cities of the Future 31
2.1 Overview of IoT and Sustainable Cities 31
2.2 Relevance of IoT to Sustainable Cities 32
2.3 Emerging Trends and Technologies in G-IoT for Smart Cities 33
2.4 Smart Development: G-IoT in Infrastructure and Urban Planning 37
2.4.1 Smart Traffic and Transportation Systems 37
2.4.2 Green Buildings and Smart Buildings 37
2.4.3 Environmental Monitoring and Urban Green Spaces 38
2.4.4 Data-Informed Urban Planning 38
2.5 G-IoT in Water and Energy Conservation Management 39
2.5.1 Smart Energy Grids 39
2.5.2 Smart Grids Powered by G-IoT: Driving Energy Efficiency, Resilience, and Sustainability 43

2.5.3 Water Conservation and Smart Water Management 44
2.6 Building Smart Communities: G-IoT for Social Sustainability and Quality of Life 47
2.7 Enhancing Urban Sustainability with G-IoT to Promote a Circular Economy 52
2.7.1 Smart Waste Management and Recycling 53
2.7.2 Organic Waste Composting (Closing the Loop on Food and Green Waste) 53
2.7.3 Digital Product Passports and Lifecycle Tracking 54
2.8 Smart Waste Management and Urban Mining 55
References 56

3 AI and the Green Internet of Things (G-IoT) in Smart Cities 57
3.1 Introduction 58
3.1.1 The Role of AI in Shaping Smart Cities 58
3.1.2 The Evolution of Smart Cities: The Role of AI 59
3.1.3 The Synergy Between AI and G-IoT 59
3.1.4 AI Applications in Smart Cities 60
3.1.5 Challenges in Implementing AI in Smart Cities 61
3.1.6 Future Prospects of AI in Smart Cities 62
3.2 AI-Driven Infrastructure and Urban Planning for Revolutionizing City Development 63
3.2.1 Introduction 63
3.2.2 The Role of AI in Urban Planning and Smart Infrastructure 65
3.2.3 The Impact of AI on Transportation and Mobility Infrastructure 65
3.2.4 Smart Infrastructure for Resilient and Disaster-Ready Cities 66
3.2.5 Challenges and Future Prospects of AI-Driven Urban Development 67
3.3 Challenges and Key Requirements for Smart Cities 68
3.3.1 Introduction 68
3.3.2 Key Challenges in Smart City Development 69
3.3.3 Key Requirements for Smart City Success 70
3.4 Smart Transportation Systems and Mobility Solutions Using AI 72
3.4.1 AI-Driven Traffic Management Systems 73
3.4.2 Autonomous Vehicles and AI-Enabled Public Transportation 74
3.4.3 Sustainable and Green Mobility Solutions 74
3.4.4 Challenges and Future Outlook 75
3.5 Leveraging AI for Sustainable Urban Living with Energy-Efficient Buildings 75
3.5.1 AI-Driven Energy Optimization in Buildings 76

3.5.2 Challenges and Future Prospects ... 78
3.6 Optimizing City Operations for Better Quality of Life with AI-Enabled Services ... 78
3.7 AI Applications in Smart Cities in the Real World ... 82
3.7.1 Energy Management and Optimization ... 83
3.7.2 Traffic and Transportation Management ... 83
3.7.3 Urban Planning and Infrastructure Development ... 84
3.7.4 Public Safety and Security ... 85
3.7.5 Water and Environmental Monitoring ... 85
3.7.6 Citizen Engagement and Services ... 86
References ... 86

4 5G Networks Towards Sustainable Smart Cities ... 89
4.1 Introduction to the Next Generation of Connectivity: 5G Technology ... 89
4.1.1 Detailed Comparison of 4G and 5G ... 90
4.1.2 New Applications Enabled by 5G ... 91
4.1.3 Specific Applications and Their Impact ... 93
4.2 Key Features and Capabilities of 5G Technology ... 94
4.2.1 Enhanced Mobile Broadband (eMBB): Driving High-Bandwidth Applications ... 95
4.2.2 Ultra-Reliable Low-Latency Communications (URLLC): Ensuring Critical Operations ... 95
4.2.3 Massive Machine-Type Communications (mMTC): Connecting the IoT Ecosystem ... 95
4.2.4 Network Slicing: Tailoring Networks for Diverse Needs ... 95
4.2.5 Beamforming and Massive MIMO: Enhancing Signal Efficiency ... 96
4.2.6 Edge Computing: Enabling Real-Time Processing ... 96
4.2.7 Energy Efficiency: Supporting Sustainable Deployments ... 96
4.2.8 Software-Defined Networking (SDN) and Network Function Virtualization (NFV): Facilitating Network Flexibility ... 96
4.3 Transforming Urban Environments Through 5G Technology ... 97
4.3.1 Enhanced Connectivity and Communication ... 97
4.3.2 Smart Infrastructure and IoT Integration ... 97
4.3.3 Intelligent Transportation Systems ... 98
4.3.4 Public Safety and Security ... 98
4.3.5 Energy Management and Environmental Sustainability ... 99
4.3.6 Healthcare Transformation ... 99
4.4 Enabling Speed, Efficiency, and Scalability by Driving 5G Networks ... 99
4.4.1 Speed: The Real-Time Revolution in Smart Cities ... 99
4.4.2 Efficiency: Optimizing Urban Resources and Connectivity ... 101
4.4.3 Scalability: Building Networks for Growing Cities ... 101

4.4.4 Governance and Policy Challenges in 5G Deployment 102
4.5 Concluding Remarks: The Role of 5G in Building Sustainable Smart Cities 103
4.5.1 5G as the Backbone of Smart Cities 103
4.5.2 Enhancing Urban Efficiency Through Network Slicing and Virtualization. 103
4.5.3 Energy Efficiency and Sustainability in Smart Cities 103
4.5.4 Accelerating the Adoption of Autonomous Vehicles and Smart Transportation 104
4.5.5 Security and Privacy Considerations in 5G-Powered Smart Cities 104
4.5.6 Challenges and Future Prospects of 5G in Smart Cities 104
References 105

5 Case Studies 107
5.1 Case Study 1: Schneider Electric's AI-Driven Smart Metering System 107
5.2 Case Study 2: Electric Vehicles and AI – Driven Sustainable Urban Mobility with Tesla 109

About the Authors

Dr. Vijay Prakash Gupta is Associate Professor and Ph.D. Coordinator at GLA University, Mathura, with 18 years of rich experience in education and research. He holds a Ph.D. in Management, MBA, MA in Economics and qualified in UGC NET in Management. Dr. Gupta has been honored by the Uttar Pradesh Higher Education Department for curriculum design under NEP-2020. He also received the 2020 PERFICIO Awards for Best Research and Innovation from BHS Foundation (New Delhi) and Campbell University (USA).

Additionally, Dr. Gupta serves as an Editor of approx. 12 books published by renowned publisher like Springer, Emerald, Wiley, Bentham Science, etc., indexed in Scopus and Web of Science and Reviewer for esteemed international journals. He has 30+ published papers in renowned national and international journals, including those indexed by ABDC, Scopus, Web of Science, and UGC CARE.

A. K. Haghi is Retired Professor and has written, co-written, edited, or co-edited more than 1000 publications, including books, book chapters, and papers in refereed journals with over 4400 citations and h-index of 35, according to the Google Scholar database. Prof. Haghi holds a B.Sc. in urban and environmental engineering from the University of North Carolina (USA) and holds two M.Sc. degrees, one in mechanical engineering from North Carolina State University (USA) and another one in applied mechanics, acoustics, and materials from the Université de Technologie de Compiègne (France). He was awarded a Ph.D. in engineering sciences at Université de Franche-Comté (France). Prof. Haghi's extensive educational background and supervisory roles underscore his expertise and contributions to the field of engineering sciences. He is appointed as Honorary Research Associate (HRA) at University of Coimbra, Portugal. He is Regular Reviewer of leading international journals.

Dr. Anuradha Yadav is Associate Professor of Marketing at DPG Degree College, Gurugram. She has more than eight years of experience in both teaching and corporate. She has publications in ABDC, Scopus and UGC care journals. She is keen Researcher with a great research interest. Consumer behaviour, Omni Channel Management, Online Marketing, Sustainability, Circular Economy, Net Zero Economy are some of her research interests.

1 Components and Requirements for Digital Infrastructure

Smart cities are reshaping urban life, leveraging digital infrastructure to improve living conditions, enhance economic efficiency, and foster sustainable development. The rapid advancements in technology, coupled with a growing awareness of environmental issues, have converged to create a new paradigm in urban planning MMCA Alumni [1]. The modern smart city is an integrated system, where digital technologies and sustainable practices combine to optimize resources, reduce waste, and improve residents' quality of life. The combination of artificial intelligence (AI) and the green internet of things (G-IoT) is essential to this shift because it allows cities to proactively respond to environmental issues and promote creativity in urban government Abraham et al. [2].

Global cities undergo a metamorphosis because of quick urban development together with modern technological changes and environmental problems. The urban population exceeds fifty percent of global citizens and researchers forecast it will reach almost seventy percent by 2050. Urban areas are encountering a multitude of difficulties with their continuous expansion that includes population congestion and road congestion and air pollution and elevated energy usage and insufficient waste control and climate alterations. The "smart city" concept represents a promising solution for dealing with current urban challenges as well as improving conditions of living for city dwellers.

Buildings which use digital systems along with data analytics together with sustainable operations work to enhance system performance and optimize resource usages and urban living conditions and eco-friendly operations. Smart cities accomplish their objective of optimal urban administration by combining several technologies, such as the Internet of Things (IoT), artificial intelligence (AI), and big data analytics [3]. The technologies are simultaneously employed to establish an environment which combines livable conditions with resiliency and forward-thinking elements. One of the most critical enablers of smart

V. P. Gupta et al., *Green IoT and AI for Sustainable Development of Smart Cities*,
Synthesis Lectures on Sustainable Development,
https://doi.org/10.1007/978-3-031-98743-4_1

cities is the development of a robust and adaptable digital infrastructure. The foundation of smart cities is digital infrastructure, which links people, systems, services, and gadgets to enable real-time monitoring, data sharing, and decision-making. Digital infrastructure is also made up of many different types of technologies, such as sensors embedded in municipal assets, communication networks, cloud computing technologies, and data centers [4]. It also allows easy data exchange among different parts of the city to enhance productivity and promote new innovations that enhance the sustainability and intelligence of urban life.

In addition to these technological innovations, sustainability is another integral aspect of successful smart cities and is increasingly important to integrate it across all aspects of urban life. The need for green, sustainable technology is particularly urgent given the increasing concern over the depletion of resources, climate change, and damage to the environment that is currently plaguing cities across the globe [5]. Adding sustainability to digital infrastructure is also critical in ensuring cities remain viable over the long term, while reducing their impact on the environment. Here the intersection of two core themes, artificial intelligence (AI) and the green internet of things (G-IoT) is central [6].

The employment of environmentally friendly IoT technology is referred to as the Green Internet of Things (G-IoT). Power consumption and power wastage the power-hungry nature of devices in non-ambient IoT deployments could significantly contribute to energy wastage. But G-IoT hopes to make these systems use less energy, helping to make them more environmentally friendly by making their IoT devices more energy-efficient. At the same time ensuring the operation benefits of IoT (real-time monitoring, smart infrastructure and resourceful solutions) are maintained and due to this the integration of cities is a substantial reduction for the carbon footprint.

"AI powers" enables cities to make smarter data-driven decisions through its life changing force working on smart city operations. AI and ML find in use scalable algorithms which can work with large amounts of IoT data for pattern-based learning and trend prediction, automated urban management and so on. AI in smart cities can be utilized for two reasons: AI is used for optimization in transportation and electricity are used for environmental surveillance and public safety are used for waste management [7]. The smart cities adopting AI are integrating G-IoT that helps to reduce their power consumption and optimize the service infrastructure. For a discussion of smart cities, a closer look at digital infrastructure that allows the functioning of smart cities is needed. The collaboration between the essential components of smart city is clarified in the chapter as description of the way their cooperation is to take place to promote development towards innovation, sustainability, and resilience. Sustainable smart urban areas are established evolving big data analytics, machine learning algorithms, the G-IoT, real time environmental monitoring, and green infrastructure solutions.

It is necessary to analyze the components and features of the digital infrastructure of smart cities, in order to optimize the full exploitation of the same. In this chapter we discuss the components that contribute to smart city infrastructure, and how they work together to enable sustainability, innovation and the resilience of urban systems. Elements—with which it is building to create an intelligent and sustainable city, and that focus on big data analytics, machine learning algorithms, the Green Internet of Things (G-IoT) real time environmental monitoring and green infrastructure solutions:

1.1 The Role of Digital Infrastructure in Smart Cities

The digital infrastructure is the soul of every smart city. "Digital infrastructure" refers to the integrated system of hardware, software, data, and communication networks—virtual and physical—that enables the gathering, processing, storage, and exchange of information in multiple urban systems. These systems, which gather data in real time on everything from garbage levels and air quality to energy use and traffic patterns, rely on sensors, actuators and networked devices. This equipment can cooperate with PC and big data through smart data management to form a "smart ecological system", and promote the level of governance, city management and decision-making of the city. Digital infrastructure is the sum of several underlying technologies that are indispensable for the functioning of the smart cities. Here are Some of the Most Important Parts:

- **Sensors and IoT Devices**: Sensors and IoT devices may be used to collect data about the physical world, such as air quality, temperature, humidity, and the number of vehicles. By connecting to these devices, which are embedded into city infrastructure such as water systems, garbage cans, traffic signals and street lights, cities can track and control resources in real time.
- **Communication networks**: Networks offer sensors, cloud platforms, and Internet of Things devices the ability to transmit and analyse data in real time. When it comes to moving data, technologies like 5G, Low Power Wide Area Networks (LPWAN) and Wi-Fi are essential to keep information moving fluidly and quickly between devices and systems all over the city.
- **Cloud Computing and Data Storage**: City officials can store and process huge amount of data created from IoT devices by using city's infrastructure by employing cloud computing. Cloud services are the fundamental premises of big data analytics, machine learning and other data-driven decision-making methods supporting smart city initiatives.
- **Big Data Analytics and AI**: The large volume of data generated by IoT devices can be utilised using data analytics and AI algorithms. Active decisions, trend forecasting, or smart operation of a city infrastructure (such as energy, waste collection, transportation, or emergency services) can be supported by these technologies.
- **Smart Governance Platforms**: The smart city is not just about technology, but also about a set of governance paradigms. (B) Digital infrastructure: The digital infrastructure also must encompass platforms for effective governance, citizen engagement and

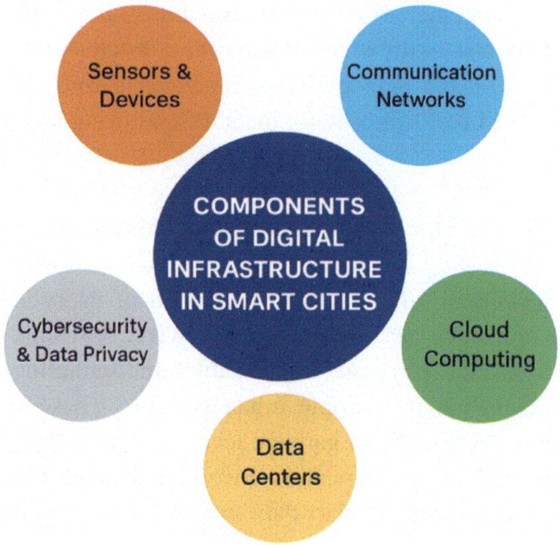

Fig. 1.1 Components of digital infrastructure in smart cities

transparency. These provide a means for local governments to access data from it's residents, interact with citizens, and to make certain that city services are distributed fairly.

- **Cybersecurity Systems**: Cybersecurity has become an important part of the digital infrastructure, given the fact that the dependence on digital technology is thriving. "Success in the connected city hinges on the safety of private data from the reach of online attackers, the sanctity of city networks, and the safeguarding of sensitive information (Fig. 1.1).

These components contribute to the support offered to the smart city infrastructures for predictive analysis, service and operation automation, and real-time management of urban systems. Digital infrastructure enables cities to reduce the consumption of power, optimise resource utilisation, minimise harm inflicted on the environment, and optimize citizens' overall quality of life.

1.1.1 Big Data and Predictive Analytics for Smart Cities

One of the most game-changing aspects smart city infrastructure would be big data! Cities generate exponentially larger volume of data as they expand and become more complicated. Every action in a city—be it in the form of traffic patterns, energy use, social media, or waste collection—generates data that can be used for the betterment of urban living. Big data analytics allows cities to analyse and interpret this wealth of information on the fly to inform decision and efficiency. One example is the application of big data to traffic monitoring, which enables dynamic traffic signal control. It can also be used to optimize public transportation routes, which would help to minimize congestion and emissions. Regarding the energy management in the city, the big data analytics can be used to monitor the electricity consumption in the city, forecast electricity demand peak and efficiently distribute electricity through smart grid. One of the greatest advantages of big data in smart cities is predictive analytics. They can use machine learning algorithms to predict future events and trends based on historical and real time data. For example, predictive models can forecast the likeliness of heatwave or heavy rainfall, 2, 43 scenario from which early warning/preparedness measures can be taken.

1.1.2 Green Internet of Things (G-IoT) and Sustainability

The concept of the Green Internet of Things (G-IoT) becomes particularly relevant in the framework of sustainable smart cities. Through the use of renewable energy and development of energy-efficient devices, G-IoT seeks to integrate sustainable development concepts into IoT systems. G-IoT aims to solve this issue through reducing the energy consumption of traditional IoT devices, with the latter being criticized for high energy usage. There are several environmental benefits of G-IoT. The G-Smart grid address power distribution over a city, and the smart streetlights minimizes electricity consumption from public lighting. Additionally, garbage collection routes may be more efficient and bin fullness can be monitored to reduce fuel consumption and decrease the environmental impact of trash disposal using G-IoT waste management systems. Other smart city systems such as G-IoT for Smart irrigation for agriculture and water saving also are part of things what we calling as G-IoT. Sensors on irrigation systems check soil moisture levels, which means farmers and people in urban areas can water less to account for overwatering. Like this smart waste management systems optimize garbage pickup routes and reduce unnecessary collection trips by using Internet of Things (IoT) devices that monitor waste levels as they fill up in real time. This also uses less energy and reduces emissions.

1.1.3 Artificial Intelligence (AI) and Machine Learning in Smart Cities

Artificial intelligence (AI): The key to smart city management one of the integral pillars of smart city management is AI. Big data can be analysed through AI algorithms—especially machine learning—to find patterns, make predictions, and even discover insights. AI is utilized in various sectors in smart cities such as public safety, healthcare, energy management, and transportation [8]. In smart cities, AI systems can help improve traffic management by processing real-time data collected from traffic sensors, cameras, and GPS devices. Instantaneous traffic signal modifications and congestion forecasting with the aid of machine learning can reduce the pollution level and congestion. Artificial intelligence (AI) may also help making energy distribution better by examining the consumption trend and predicting the demand rise, so that energy may be used in a more efficient manner. Medical Artificial Intelligence (MAI) can capture citizens' health-related data can anticipate public health emergencies and issue Medical advice to citizens [9]. AI can be used for public safety too: Its video analytics can be used to assess video surveillance feeds, detect suspicious activity, and inform law enforcement immediately.

1.1.4 The Future of Digital Infrastructure for Smart Cities

The digital infrastructure that supports cities needs to evolve with them. Enabling faster and reliable device-to-device interaction; boosting data security; and enabling real-time decision-making at the edge of the network, information and communication technologies including 5G, blockchain technology and edge computing will greatly extend the capabilities of smart cities [10]. These innovations will also help to ensure smart cities are more resilient, efficient and sustainable in the process.

As cities respond to the demands of growing populations, the threat of climate change and sustainability goals, digital infrastructure will continue to grow in importance in the future. Through the strategic investment in strong, flexible digital infrastructure we can create technically advanced, environmentally responsible and equitable cities. Ultimately, the degree to which digital technology, sustainable practices, and creative models of governance are integrated in harmony, producing smarter and more resilient cities, will influence how our urban environments will grow in the days to come.

This chapter looks at the components and necessary conditions for building digital infrastructure that supports sustainable smart cities. By analyzing their integration we investigate how AI and G-IoT are merging with machine learning algorithms in order to change the nature of urban management. Furthermore, this chapter looks at how cities are deploying green infrastructure to mitigate the impacts of climate change, the role of G-IoT in enabling energy-efficient IoT devices, and the importance of big data for the intelligent management of urban systems. Drawing on inter-disciplinary approaches and practical examples we illuminate how new technologies are enabling innovative solutions to meet the challenge of sustainable development.

1.2 Big Data for a Sustainable Smart City

1.2.1 The Role of Big Data in Smart Cities

Big data is essential to turning cities into smart, sustainable places. Cities now have more immediate access than ever before to real-time data about nearly everything that happens in urban areas, through the widespread new linkages—the so-called Internet of Things (IoT)—that sensors, cameras, and other connected technology has allowed them to create. Big data helps city leaders make more evidence-based choices to support sustainability and efficiency—from monitoring garbage collection and public health, to measuring traffic bottlenecks and energy consumption. With respect to urban management, big data can help optimize resource allocation, predict demand, and identify inefficiencies. Cities, for example, can use smart meter data to monitor energy use and detect anomalies that signal dramatic spikes in demand—evidence of equipment failure or inefficiency. By tracking traffic data, cities can improve timing of traffic signals, reduce congestion and reduce emissions. Similarly, trash management systems could be used to optimize garbage collection routes and monitor levels in garbage cans via sensors, reducing fuel use and the carbon footprint of the city. The concept of sustainability is closely related with the idea of a smart city. Digital technology is used by smart cities to enhance the quality of life of their citizens and reduce the environmental footprint of city operations. 3 Big Data, the large and heterogeneous data produced by urban systems, people, and gadgets is one of the major contributors to make a smart city a reality. However, if the data is efficiently collected, analyzed and utilized, it can help to inform resource use, governance, urban planning, and environmental sustainability. Big Data allows cities to make decisions that are more data driven, which in turn reduces the need for resources, provides better service, and ultimately makes life in cities better for people. Big data can be put to use helping cities to intelligently improve their systems for sustainability and efficiency, whether in trash disposal and water management or energy use and transportation networks. Predictive analytics and real-time monitoring can help predict and prevent problems such as traffic jams, energy-gluts, pollution, and overflowing garbage, all of which are crucial to sustainable urban development (Fig. 1.2).

This section focuses on the role of big data in building sustainable smart cities. We'll also consider how Big Data can be used to enhance environmental quality, urban sustainability, and management of natural resources. In this way, we will show that Big Data, along with emerging technologies such as the Internet of Things (IoT), artificial intelligence (AI) and machine-learning applications, can innovate urban systems to become more green and sustainable.

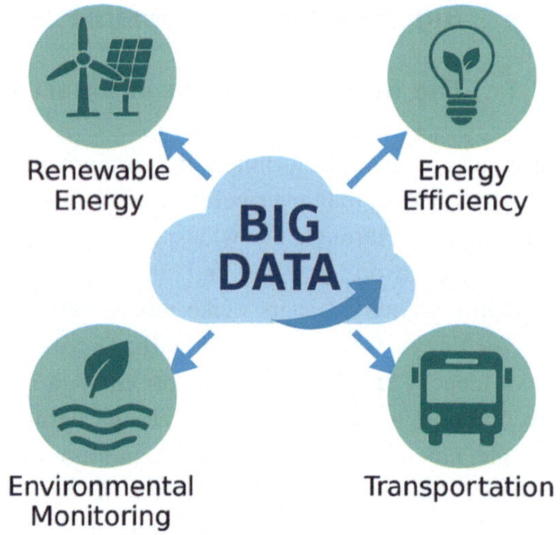

Fig. 1.2 Big data for sustainable smart cities

1.2.2 The Importance of Predictive Analytics

Predictive analytics is also among the most important values of big data in smart cities. Machine learning systems can recognize patterns and forecast future events by looking at history as well as what's happening today. For example, reducing congestion, cities may change traffic light timings on-the-fly when predictive software predicts traffic jams from historic traffic data, weather or public events. Predictive analytics can certainly aid disaster preparedness for environmental sustainability. Cities can prepare for heat waves, downpours or floods by analyzing climate data. This allows them to take action, for example issuing public warnings, or changing infrastructure to reduce potential damage. With big data and predictive analytics, cities can become more prepared for coming challenges, also develop a more resilient and sustainable urban future [11].

1.2.3 The Role of Big Data in Smart Cities

Knowing the numerous sources where this data collects is an essential of an understanding big data for smart cities. Smart city data: In addition to conventional urban data (such as traffic flows or energy use), a smart city generates data from sensors, the IoT, social media, and other sources. Better decisions become feasible due to the full picture of urban life that these information sources enable.

IoT and Real-Time Data Gathering: To gather the data in real time, interconnected sensors, cameras, and other devices are utilized by smart cities. Such devices are part of public services (think of garbage collection, water networks and public transportation systems) and even infrastructure (buildings, roads, bridges and streetlights). These sensors offer a never-ending source of data, monitoring such variables as the amount of trash, the amount of energy being used, the temperature, the movement of traffic and the quality of the air.

Data From Public Services: Vast amount of data is generated from the systems in towns like garbage collection system, water treatment plant and electrical grid system. Its data can be utilized to improve effectiveness and resource allocation. In order to avoid waste of energy, smart grids, for example, can automatically adjust distribution and monitor power usage in communities.

Social Media and Open Data portables: Open government data and data coming from social media also provides a contribution to the Big Data world along with data from sensors. For instance, social media feeds include current data applicable to events, public sentiment, transportation issues, and environmental conditions. Open data platforms, on the contrary, open up urban data to the public, and encourage citizens to be more active in the planning and decision making processes of the municipality.

1.2.4 Big Data for Environmental Sustainability

Cities around the world are struggling with environmental problems arising from increasing urbanization, climate change and resource shortages. With the global population increasingly moving to urban areas, environmental problems, such as pollution, solid waste, energy consumption and water scarcity, have become more significant. In this regard, Big Data provides essential weapons to improve environmental sustainability through knowledge-driven decision-making and efficient resources utilization.

- **Reducing Energy Consumption**

Energy consumption one of the most significant of the areas that Big Data might have an impact on is the consumption of energy. Cities gobble up a lot of energy, much of which ends up wasted due to inefficient systems, aging infrastructure and poor design. Big Data will allow cities to better manage energy, reduce energy consumption and move on to renewable sources. Smart grids and energy management solutions are a significant part in this effort. Smart grids use sensors, data analytics and AI to measure and manage power usage in real time. Based on power consumption behavior data a smart network can adjust the available power supply in response to fluctuations in consumption, this improves energy efficiency. For instance, when energy demands rise, the grid could immediately know which areas require more power and deliver it accordingly. Likewise, in periods of lower demand, surplus power can be stored or pumped into renewable sources—lessening reliance on fossil fuel. And the Big Data analytics can be used to predict energy demand

and optimize the consumption pattern in many industries (along with Food and Industry 5.0 (2024)). For example, predictive appliances could predict peak hours, and adjust the use of air conditioning, heating, or lighting facilities in homes or commercial spaces. By avoiding wasteful energy consumption, this type of automation not only reduces energy costs, but reduces the environmental footprint.

- **Waste Management and Resource Recycling**

Cities are now grappling with the garbage management problems resulting from a growing population and scant landfill space. The global move towards more circular economy principles (where products and resources are reused, repaired, recycled, or re-manufactured) has emphasised the importance of good waste management. Connected big data, when combined with IoT sensors, provides real-time information on trash generation and collection, enabling cities to optimize their waste handling and improve their recycling rates. Smart waste management solutions make use of data produced by the IoT sensors installed in garbage bins and dumpsters to measure the fullness of the trash. Information about when bins are actually full or close to reaching capacity is used for manipulation of garbage collection schedules, ultimately reducing fuel costs, emissions, and overflow. Furthermore, communities could use this information to map where trash is being generated in different areas and enact interventions like offering more bins, or promoting recycling in high waste zones. Aside from garbage collection, Big Data can also be involved with recycling resources. For example, robust ML algorithms could evaluate home recycling habits, identify trends in trash creation, and suggest ways to reduce waste or recycle more. This data can be used for educating communities on how to recycle more efficiently and creating incentive-driven programs to stop generating trash.

- **Air Quality Monitoring and Pollution Control**

Air quality is an important indicator of the environmental sustainability of the city. Bad air quality—influenced by the likes of traffic pollutants, industrial pollutants and construction—can manifest in a number of health issues as well as contributing to climate change. Hence an interest, by city planners and politicians, in monitoring and enhancing urban air quality. Real-time monitoring of air quality from Big Data because sensors are being used to measure gases such as NO_2, CO or PM and ozone, real-time air quality monitoring is possible in Big Data. By gathering this information, cities can find pollution hotspots, track changes in air quality and intervene. As, for instance, when pollution levels pass a harmful threshold in a specific area, real-time data could lead to actions such as turning on air purifiers, issuing public health advisories or restricting vehicle access to limit emissions. Big Data can even be leveraged to predict air quality trends through weather conditions, traffic flow and other contributing factors. Artificial intelligence algorithms can interpret a history of air-quality data to predict levels of pollution, providing critical information to local officials. This lets towns anticipate air quality problems and implement smart solutions, such as shifting transportation temporarily or shutting down industry when there are high levels of pollution.

- **Water Management**

Water scarcity is a major problem throughout the world, and cities are increasingly feeling the heat to use their water resources properly. Big Data could help urban water systems cut waste, improve efficiency and ensure that water reaches the people who need it most. Sensors connected to monitors monitor water usage, consumption patterns, leaks and reservoir levels, which can help manage water on smart water management systems. This data is analyzed to identify inefficiencies and prevent loss of water. For example, IoT sensors are used in leak detection systems in water pipelines to identify leaks or bursts, to send notification to concerned authorities for taking immediate actions to prevent waste of water. Predictive water management is also supported by Big Data. Through analysis of water use and weather, predictive models can forecast periods of peak demand or impending drought. These findings could factor into water conservation efforts, including altering irrigation timings, encouraging water-efficient appliances, and a better allocation of water between communities.

1.2.5 Big Data for Urban Mobility and Transportation

Transportation networks are fundamental in the operation of a smart city, and they can have a direct impact on sustainability. Long queues, long travel times and inefficient transports networks generate environmental pressures (such as excessive fuel use and air pollution) as well as economic inefficiencies. Urban mobility Big Data contributes significantly to the improvement of mobility in city environments, influencing traffic control and public transportation improvement, as well as the adoption of new sustainable transports.

Traffic: Traffic patterns can be monitored and congestion areas detected by making use of real-time data from traffic sensors, GPS-based devices, and cameras. This information can be exploited to change signal lights and regulate road usage, thus dynamically rerouting automobiles in order to minimize congestion. AI programs can predict traffic jams and take steps to avert them, easing the flow of cars as well as the man-made clouds of carbon created by idling vehicles.

Public Transportation: Big Data may optimize public transportation including passenger data analysis, routes and schedules. For example, cities can use Big Data to identify underperforming bus routes, and then adjust frequency or reroute buses to serve high-demand areas more effectively. What's more, predictive analytics would be able to provide real-time data to help keep passengers informed of delays or the best route to take, acting to keep public transport an attractive and reliable option.

Shared Mobility and Electric Cars (EVs): Big Data also supports the evolution of shared mobility services (example- ride-hailing apps and bike sharing) as well as electric cars (EVs). For example, through the analysis of mobility patterns, it generates better locations to create electric vehicle charging points, helping to track the availability of a fleet or supporting the use of shared transportation that reduces the number of private vehicles on the road, which means less land and an emission reduction are the direct and significant effects of this."

Conclusion

Sustainable smart cities are built on the foundation of Big Data. It enables cities to address some of the pressing environmental and resource issues by maximizing resource utilization, improving the performance of urban systems, and enabling informed decision making. By monitoring energy use and rubbish collection, to air quality and water conservation, Big Data can give many of our cities the information that they need to build green, resilient and liveable urban spaces. By bringing together IoT, AI and machine learning, Big Data offers new possibilities for cities to drive sustainability in every part of the urban environment. With more municipalities worldwide pursuing growth and facing new environmental challenges, Big Data will inevitably become more of a key factor not only in the construction of smart cities, but the construction of sustainable our sustainable urban societies—for the sake of present as well as future residents.

1.3 G-IoT for Sustainable Smart Cities and Society

1.3.1 Defining Green Internet of Things (G-IoT)

The Green IoT (G-IoT) is one of the recent advancements in the ecosystem of smart city that brings in the potential for improving the energy efficiency of the things in the IoT. Existing IoT systems are often energy-hungry as they require continuous communication between various devices and centralised (cloud) servers. This perpetual data transmission obviously entails further energy consumption, which goes against the expectations for sustainable urban districts. G-IoT addresses this challenge by focusing on minimizing the power consumption of IoT devices. This can be achieved by various methods such as low-power sensors, energy-efficient communication protocols, and energy harvesting implementations. G-IoT devices are able to utilize techniques, such as adaptive data transmission, in which devices only send data when necessary or when significant changes are sensed, minimising the energy consumption. Introduction A "smart city" is a city that is equipped with digital technologies that result in efficient and productive use of resources, while also improving the quality of life of those living in the city, in addition to other operational improvements. With technology embedded ubiquitously into services and infrastructure the city smart dreams to provide for "comfort, convenience, and choices and better opportunities for all citizens, whether citizens, workers or tourists" [12]. The key reason for interest in the topic is the possibility that such technologies could transform urban life to be healthier, more livable, and more environmentally sustainable. In this chapter, we emphasize the integration of the Green Internet of Things (G-IoT) with Artificial Intelligence (AI) for building green and sustainable smart cities. The trends in this integration demonstrate the way smart city automation can be scaled up to offer significant environmental and operational benefits.

The Potential of G-IoT and AI in Upgrading Urban Life

The fusion of G-IoT with AI in smart cities indeed holds the aforesaid excellent promises that would transform the overall urban life-styled. Real-time data collection and analysis allows G-IoT devices to communicate with AI systems to optimize city management practices. Transport can dynamically adjust to reduce congestion and pollution, while AI can predict maintenance needs for infrastructure to keep things running smoothly. G-IoT and AI-powered smart city projects can track energy and water consumption and air quality instantly to take immediate corrective actions. That creates better space for residents, and helps governments achieve sustainability goals.

The Convergence of AI and G-IoT with Machine Learning Algorithms

The amalgamation of AI and G-IoT with machine learning models provides new avenues in management of smart cities. Predictive analytics enabled by machine learning can foresee urban problems and solve them as per the requirements [13]. For example, waste collection systems connected to G-IoT sensors are able to monitor the level of garbage and predict optimal times for disposal, reducing both expenses and the environmental footprint. AI-driven automatization could also facilitate urban planning. Algorithms can process enormous amounts of data to detect trends, like population growth or shifting commute habits. This information could be used to support more efficient transportation systems, or improve land use in rapidly urbanizing areas.

Energy Efficiency with G-IoT

The Green Internet of Things is also able to reduce energy consumption. G-IoT is supposed to work with low power consumption which is also important to achieve energy efficiency in urban environment. Smart lighting systems, for instance, utilize G-IoT technology for dimming lights based on the surrounding light intensity and the presence of human beings, in order to conserve power. Furthermore, G-IoT setups could ramp up renewable energy utilization. Solar panels, wind turbines, and other renewable energy systems integrated with the IoT network can help communities keep a closer eye on and manage the amount of energy being produced and wasted, all the while reducing the need for energy derived from non-renewable sources (Fig. 1.3).

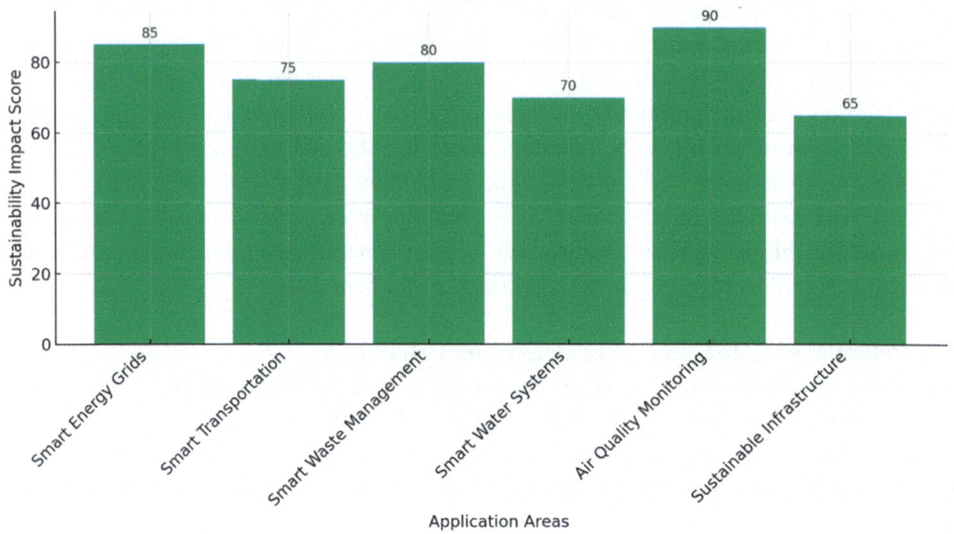

Fig. 1.3 G-IoT for sustainable smart cities and society

This chart illustrates the sustainability impact of Green IoT (G-IoT) across key smart city domains. Highest impact is seen in energy management and smart transportation, reflecting G-IoT's role in reducing emissions and enhancing efficiency. Moderate influence is observed in healthcare, water conservation, and waste management, highlighting its broader societal benefits.

1.3.2 Practical Applications Through Case Studies

This book provides practical examples with real world case studies on how smart cities worldwide are using G-IoT and AI technology effectively. For instance:

- **Barcelona, Spain**: The city's smart water systems use IoT sensors to monitor leaks and improve water distribution, saving millions of liters annually.
- **Singapore**: AI-based monitoring systems with G-IoT devices enhances public safety and efficient traffic management.
- **Amsterdam, Netherlands**: The smart energy meters used in the city are dedicated to mitigating sustainable energy usage and measuring the real-time consumption. These are exemplary cases showing the transformational impact of G-IoT and AI integration in urban systems, and they highlight the relevancy and the scalability.

1.3.3 Interdisciplinary Ideas for Smart City Development

Smart cities creation are operating across different disciplines that encompasses technological skills, urban planning practices, environmental studies and public policy as well [14]. G-IoT and AI are driving forces behind the creation of a sustainable ecosystem that address a wide range of issues ranging from environmental protection to resilient infrastructure. For instance, interdisciplinary collaboration between engineers and environmental scientists has resulted in smart sensors which can measure pollution and recommend practical steps. Likewise, liaisons between technologists and policymakers ensure that solutions for smart cities are aligned with the legal framework and social requirements.

1.3.4 Integrating Technology and Sustainability Concepts

Smart city initiatives intersect technology and sustainable goals. By combining these principles, G-IoT and AI bring the parties involved one-step closer to achieving common aims. For example:

- **Sustainable**: AI can analyse data from G-IoT devices to provide intelligence driven insights, such as avoidable energy waste or maximum trash recycling.
- **Community engagement**: Smart city platforms allow people to take part in decision making and to create a sense of ownership and accountability.
- **Economic development**: The deployment of G-IoT and AI will foster economic growth by creating employment opportunities (technology development and city management jobs).

1.3.5 The Role of G-IoT in Smart City Sustainability

So in this context, the combination of G-IoT technologies is one of the critical tasks in the long run for the smart city environment. Through reducing the energy consumption of IoT devices, the carbon footprint of the entire city could be significantly reduced. In addition, G-IoT systems are often integrated with renewable energy sources like solar panels or wind generators to reduce the relaince on traditional carbon-heavy energy sources. One of the most successful examples of G-IoT is the roll out of smart street lighting. Standard streetlights shine all night long, whether a lot or little traffic is on the road. "By comparison G-IoT smart streetlights adjust light levels based on current data from sensors that detect the presence of people or cars. It is an energy-saving as well as environmentally friendly solution that will also help in reducing the amount of light pollution in cities.

1.3.6 Case Studies: G-IoT in Action

City of Copenhagen1 in Denmark uses G-IoT solutions to monitoring energy consumption in buildings, streetlights and public transportation. The city replaced all streetlights with energy-efficient smart technology, solar-powered sensors that adjust the intensity of lights based activity by pedestrians. This has led in turn to great reduction in energy consumption and even a lower carbon footprint for the city. A third example would be the 'smart grid' technologies that are being tested in locations including San Diego, US. G-IoT is adopted by smart grids to monitor energy consumption and also to deliver the electricity in an efficient way over the grid. These networks use sensors to detect changes in energy consumption and adjust power flow accordingly, cutting power losses and increasing grid resilience.

1.4 Innovative Approaches in Environmental Monitoring

1.4.1 Real-Time Environmental Monitoring

Environmental surveillance is an inherent part of smart city, especially for the sustainability city. By deploying IoT sensors throughout cities, municipalities can keep a constant eye on environmental factors like the presence of pollution, quality of the water, temperature, and noise levels. These live data feeds enable cities to act more swiftly in response to mounting environmental threats, such as air pollution, taints in the water or heat waves. For instance, systems for monitoring air quality in real time may measure the level of pollutants, such as particulate matter (PM), nitrogen dioxide (NO_2), and carbon monoxide (CO). When pollution drops to safe levels, cities can then issue public health alerts or implement traffic bans or turn on air purifiers in affected areas. Likewise, water quality sensors could be used to detect pollution at rivers, lakes or other sources of drinking water, providing authorities with a way to respond quickly to suspected public health threats. Environmental monitoring is the collection of systematic data to study and ascertain the status of our environment. With concerns about the climate change, pollution and the loss of biodiversity mounting, creative approaches toward environmental monitoring have become critical. Traditional methods are resource intensive and lack scalability, even though important. New technological advances, for example networks of sensors, satellite maps, and data-centered monitoring systems are transforming how we see and respond to environmental problems. This chapter presents, these new approaches, and highlights their importance in promoting sustainable environmental management.

1.4.2 The Role of IoT in Environmental Monitoring

The IoT has reshaped environmental surveillance with real-time data collection and analysis functionalities. IoT systems create a web of sensors that monitors, for example, air quality, wáter levels, pore health and temperature. These methods provide accurate, local, and successive information, indispensable for early acts.

- **Smart Air Quality Monitoring**: The IoT system with pollution sensors monitors air quality in urban as well as industrial areas. Urban areas such as of Beijing and New Delhi use IoT networks to identify PM2.5, PM10 and make informed decisions to reduce air pollution.
- **Water Resource Management**: Smart sensors that report back water quality to rivers, lakes, and reservoirs, can detect contaminants, check water levels, and ensure equal distribution rates for household and agricultural use. For example, intelligent irrigation technologies use soil moisture information to minimize water usage in drought regions.

1.4.3 Remote Sensing and Satellite Imagery

Remote sensing has emerged as an field important for environmental assessment. Modern sensor-equipped satellites take high resolution photographs and data of the earth's surface, and the data can be used for large-scale environmental assessment.

- **Deforestation**: Satellite imagery has been widely used to measure deforestation and land use change. Technology, such as Google Earth Engine, make it possible for researchers to monitor forest cover through time and to identify hotspots of illegal timber cutting.
- **Climate Change**: Satellites keep watch on important climate indicators, including melting glaciers, rising sea levels, where greenhouse gases are in high concentration. NASA's Earth Observing System also provides critical information for global climate research.
- **Biodiversity Mapping**: Remote sensing plays a key role in mapping ecosystems and tracking animal migration, allowing conservationists to protect threatened species and landscapes in a cost-effective manner.

1.4.4 Artificial Intelligence and Big Data Analytics

There are new avenues created by the fusion of AI with big data analysis in environmental monitoring. AI systems sift through massive amounts of data to find patterns, make predictions and automate decisions.

- **Disaster Management Predictive Analytics**: AI algorithms analyse historical information on weather patterns, earthquake movements and water flows to predict and manage natural calamities. AI-powered early warning systems save lives and avert economic losses.
- **Pollution**: AI-based analytics identify pollution sources, enhance waste-management systems and provide regulatory tools. Abnormalities in environmental data can, for example, be detected by machine learning algorithms, which may point at issues.
- **Animal Conservation**: AI-powered systems scan camera trap photos, acoustic recordings, and drone video to track and count animals, and detect illegal poaching operations.

1.4.5 Citizen Science and Crowdsourcing

Citizen science initiatives have emerged as a tool for environmental monitoring. By engaging the public, these projects harness the power of team work to collect and process data in massive amounts.

- **Air and Water Quality Monitoring**: Low-cost sensor kits allow anyone to test the local air and water quality, sending data to a central source. For example, the "Air Quality Egg" project provides a way for citizens to measure the air pollution in their localities.
- **Biodiversity Observation**: With applications like iNaturalist and eBird, people are encouraged to register observations of plants and animals, producing large databases that academics can use to analyze trends in biodiversity.
- **Reporting Plastic Pollution**: Volunteers around the world use smartphone apps to report plastic garbage washed up on shores or found at sea, data that are critical for clean-up efforts and policy advocacy.

1.4.6 Drones and Unmanned Aerial Vehicles (UAVs)

Drones have become an essential tool for environmental monitoring from difficult-to-reach locations, as they can capture high resolution data. UAVs are used in various applications, including:

- **Forest Monitoring**: Drones monitor forests to assess tree health, measure biomass and spot illegal activities such as logging and mining.
- **Coastal and Marine Monitoring**: Drones use their surveillance to watch coral reefs, follow marine wildlife and map coastal erosion. It provides holistic perspective, which is difficult to achieve with traditional techniques.

- **Disaster Assessment**: In the wake of natural disasters, drones capture live video of disaster-stricken areas to support damage assessment and relief planning.

1.4.7 Blockchain for Environmental Data Integrity

Blockchain ensures that environmental data is transparent and trustworthy. Through creating immutable data records, blockchain addresses issues related to data tampering and to trust among partners in the value chain.

- **Carbon Credit Trading**: Decentralized ledgers enable clear visibility into carbon credits and responsibility for any carbon offset program.
- **Supply Chain Tracking**: Blockchain traces an item's ecological impact from creation to disposal, supporting green standards in the enterprise.
- **Water Management**: The technology enables a de-centralized water trade system and ensures equitable distribution, thereby preventing conflicts on shared resources.

1.4.8 Challenges and Future Directions

However, many barriers remain in environmental monitoring. Modern technology is still in play, if not held back by high prices, data privacy concerns and the digital divide. The generated data size is also too big and requires a robust facility for the data storage processing and analysis. It is expected that future advances in environmental monitoring will combine different technologies. Such a hybrid device that combines IoT, AI, and blockchain should provide full capable and trustworthy support regardless of the above environment. International collaboration and the political will of governments will also be crucial in taking these technologies global (Table 1.1).

1.4.9 The Use of AI in Environmental Monitoring

AI in Environmental monitoring Artificial Intelligence has a high potential in Environmental monitoring. AI algorithms can analyze massive flows of environmental data to detect patterns, predict trends and make decisions automatically. For example, learning methods can be used for examining historical and current air quality data and forecasting pollution levels mm h ahead, mm d into the future, so that local authorities could respond in advance. Similarly, AI-based systems can also help optimize resource utilization around environmental topics. For example, if a wildfire starts near a city, drones and sensors equipped with artificial intelligence might detect the environmental impact, like decaying air quality, and then notify residents and deploy firefighters to be on hand to

Table 1.1 Urban challenges and smart city solutions

Urban challenge	Smart city solution	Impact
Traffic congestion	AI-based traffic management systems	Reduced travel delays by up to 25% in Singapore
Air pollution	AI-optimized traffic signals (e.g., Google's Project Green Light)	Decreased stops by 30% and emissions by 10% in pilot cities
Waste management	AI-powered smart bins and route optimization	Increased collection efficiency by 30% in Shanghai
Energy inefficiency	AI-driven smart grids	Reduced energy consumption by 25% in smart buildings
Water scarcity	AI-assisted water management systems	Achieved 15% reduction in water wastage

limit damage. When AI and environmental surveillance are put together this way, maybe we can know which cities are going to need help most urgently, and prevent a bit more catastrophic damage.

1.5 Green Infrastructure and Urban Adaptation

1.5.1 Green Infrastructure for Sustainable Cities

Green infrastructure is a system of natural and semi-natural systems designed to reduce storm water, cool urban heat islands, increase biodiversity, and improve air and water quality. Green infrastructure, in contrast to traditional "gray" infrastructure that is made from concrete and steel, uses natural processes to perform, often more cost effectively, functions such as stormwater capture, pollution reduction, and habitat enhancement. Green infrastructure reduces air pollution (for example, by green roofs that provide insulation and retain rainfall), increases biodiversity (as in urban trees) and directs rainwater through permeable pavements into the ground instead of the stormwater system. Urban parks and green corridors are also part of green infrastructure, they provide spaces for recreational activities by citizens as well as contribute to mental health and well-being. Cities across the globe are experiencing very significant challenges related to climate change, degradation of the environment and rapid population growth. To overcome such challenges, cities are increasingly relying on green infrastructure as a sustainable model of urban adaptation. Green infrastructure are natural and semi-natural systems that provide ecological, social and economic benefits by integrating nature and nature based solutions into urban areas. These may ranges from Parks, Green roof, pervious pavement to wetlands. This chapter examines the role of green infrastructure in building resilient cities, its benefits, and emerging approaches to its delivery.

1.5.2 The Role of Green Infrastructure in Urban Adaptation

Green infrastructure is now recognized as having a critical role in helping cities to cope with climate change and other environmental challenges. Emulating natural ecosystems, it provides sustainable solutions for managing urban problems like flooding, heat islands, and air pollution.

- Stormwater Management: Green infrastructure helps control urban flooding by capturing and filtering storm water. (Urban projects with features like rain gardens, bioswales, and wetland areas, would capture runoff, reducing waterlogging and lessening the pressure on drainage systems.)
- Urban Heat Island Mitigation: Learn how green roofs and tree canopies lower surface temperatures to reduce the impact of the urban heat island. Cooler and greener areas enhance the thermal comfort of occupants.
- Clean Air: Trees and plants in urban areas clean the air y removing pollutants. Trees and vegetation are carbon sponges, absorbing carbon dioxide and aiding climate change solutions.
- Habitat-creation: Green infrastructure adds to species' habitats particularly for densely populated city areas.

1.5.3 Benefits of Green Infrastructure

The use of green infrastructure provides multiple benefits, not only to the environment. This combination of advantages renders it an important part of urban adaption strategies.

- Environmental Value: Green infrastructure provides ecosystem services to a community, including water quality improvement, soil stabilization, and carbon capture." It also contributes to climate control and reduces greenhouse gas emissions.
- Social Good: Green space is good for your mental and physical health, encourages community cohesiveness, and enhances quality of life. Research has indicated that urban greenery reduces stress and increases physical activity.
- Economic advantages: Green infrastructure reduces the costs of infrastructure—since green infrastructure does not generate profit, there are no additional costs. For example, permeable pavements reduce costs of stormwater controls.
- Climate Change Resistance: women infrastructure increases a city's resistance to extreme weather conditions, like flooding, droughts and heat waves, by reinforcing natural systems.

1.5.4 Innovative Strategies for Green Infrastructure Implementation

Sweet Cities around the world have also redoubled their efforts to produce green infrastructure creatively. These manoeuvres integrate traditional methods with new technology and interdisciplinary cooperation.

- Green Roofs and Walls: For both us and the environment, plants on roofs and along walls are highly beneficial for anyone living pretty much anywhere in the world. Random example: the isle of Singapore's Marina Barrage. These roofs lower the energy used by buildings and improve insulation. Vertical green walls provide similar benefits and add an urban aesthetic.
- Urban Forests: Efforts such as Tokyo's "Midorinokaze" campaign focus on dropping trees in urban spaces as a way to encourage growth and clean up the air.
- Smart Water Management: Tools such as IoT-based sensors can monitor stormwater systems and optimize water retention and use. Smart technology is used to handle rainfall in green infrastructural systems in Copenhagen, for instance.
- Community-driven approaches: Involvement of local residents in green infrastructure planning and upkeep ensures its maintenance. There are programs such as New York City's "Greenstreets" that make unused spots bright green.
- Public–private Partnerships (PPPs): Working together, governments and the business have access to funding and expertise for large-scale projects of green infrastructure. For example, the "Thames Tideway Tunnel" in London combines nature-based solutions with engineering to control rainwater.

1.5.5 Case Studies of Green Infrastructure Success

The examples below demonstrate how cities have successfully used green infrastructure as a way to address urban issues:

- Rotterdam, the Netherlands: The city's water squares double as public spaces during the dry seasons, and as stormwater retention basins after heavy rainfall to help prevent urban flooding.
- Melbourne, Australia: A tree canopy coverage plan in the Urban Forest Strategy aims to reduce heat stress and increase biodiversity in the city.
- Bogotá, Colombia: In the "Green Corridors" project the connection between parks is intensified, creating a network of green areas, essential for human and ecological health.

- Seoul, South Korea: Restoring the city's Cheonggyecheon Stream transformed a sewage-ridden waterway into a lively green corridor, stimulating tourism and reducing urban heat.

1.5.6 Challenges in Implementing Green Infrastructure

Green infrastructure brings with it a series of benefits but it also raises issues. These include:

- Space constraints: Urban areas are often not well equipped to support large green projects. Unconventional design, including vertical gardens, help to solve this problem.
- Fiscal Challenges: Green infrastructure requires a significant up-front expenditure. Nevertheless, cost sharing, sponsorship, and PPP may play a role to soften financial burden.
- Maintenance: Regular maintenance is key to the success of green infrastructure over the long term. Community involvement and high-tech monitoring systems mean less trouble with maintenance.
- Materiel Barriers: Legacy urban design regulations may hinder implementation of green infrastructure. Policymakers should support sustainable urban development plans.

1.5.7 The Future of Green Infrastructure

The fate of green infrastructure relies on the convergence of cutting-edge technology and cross-sectoral cooperation. New developments and emerging approaches include: •

- Nature-Based Solutions (NBS): Integrating healthy ecosystems and novel engineering solutions for sustainable urban development.
- Digital Twin Models: Utilizing virtual environments to design and better predict the outcomes of green infrastructure projects in the United States.
- Carbon–Neutral Cities: Building an urban infrastructure that's green enough to cancel out its carbon pollution.
- Citizen Science: Enabling residents to track and participate in green infrastructure projects through apps and web portals.

1.5.8 Urban Adaptation to Climate Change

Cities struggling to deal with the growing impacts of climate change have a powerful tool at their disposal, but they may not even know it. Around the world, cities are starting to embrace nature-based solutions to manage key environmental issues like flooding, heatwaves and extreme weather events. For instance, the "Urban Forest Strategy" of the city of Melbourne, Australia seeks to increase the tree canopy cover of the city, in order to combat to the urban heat island effect. In planting more trees and creating green spaces, Melbourne is not only reducing its energy consumption, but also is creating more comfortable environments for the people who live there, a particularly poignant fact in low-income areas, which are most susceptible to heatwaves.

1.5.9 Case Study: The "Sponge City" Concept

In China, meanwhile, the concept of "sponge cities" is being championed as a way to relieve urban flooding and water scarcity. Sponge cities are supposed to soak up and retain rain water through the deployment of permeable surfaces, green roofs, and systems that collect rainwater. Not only do these cities handle stormwater better, but they also reduce the risk of floods and help save water for use down the road. The green infrastructure projects in Wuhan, e.g., water plazas, green roofs, and wetlands are helping to create urban places more resilient and less vulnerable than before.

1.6 Smart Cities Management and Innovations for Sustainable Development

1.6.1 AI and Machine Learning in Smart Cities Management

Smart city management is being revolutionized by Artificial Intelligence (AI) and machine learning [15]. These technologies permit cities to sift through endless data in real time and make decisions for better urban processes. Machine learning systems, for example, might recognize patterns in traffic data, minimize energy consumption, and improve waste management. In energy management, AI can predict fluctuations in demand, enabling energy suppliers to adjust supply in real time and prevent waste. And in transportation, AI systems could analyze traffic data in real time to speed traffic flow, reduce congestion and reduce emissions. Machine learning models can also be used for preventive maintenance, identifying potential problems in a city's infrastructure before they turn into costly breakdowns.

Table 1.2 Global smart city rankings by digital infrastructure readiness (2024)

City	Digital readiness score (out of 100)	Key infrastructure highlight
Singapore	92	Real-time surveillance, smart transport integration
Zurich	89	Renewable-powered grid with AI automation
Oslo	87	IoT water systems, green tech integration
Tokyo	85	AI-powered metro and energy-efficient buildings
New York	82	Smart grids, AI-based traffic optimization

Source IMD Smart City Index 2024

1.6.2 Innovations in Smart City Governance

Smart Cities require new models of governance that combine technology, sustainability and social equity. And data-driven decision-making empowered by AI and big data analytics assists city leaders in identifying programs that have the greatest impact on people' lives. Additionally, new patterns of governance that are more transparent and involve the community are emerging. A number of smart cities already offer online portals through which citizens can file complaints, track the progress of burgeoning projects and be engaged in the decision-making process. From cities such as New York and London, open data campaigns represent one of the leading models of creative governance. Such systems make large troves of municipal data accessible to the public—from transit timetables to environmental data—providing an opportunity for people, corporations and researchers to study patterns and propose solutions (Table 1.2).

1.6.3 Integrating Technology and Sustainability

Smart cities: It's key to combine technology with sustainable design. Through the combination of digital infrastructure and green technology, cities can address the problem of increasing urbanization, and at the same time minimize their environmental impact. Innovations in transport, energy and rubbish means that even cities with high living standards can reduce their carbon emissions. Smart cities isn't just a matter of technology, but about creating more sustainable, resilient and efficient environments in our urban spaces. City services could be improved, environmental damages reduced, and citizens' well-being enhanced by means of merged AI, G-IoT, big data and green infrastructure. In practical terms and with the help of the multidisciplinary point of view followed in this chapter, we will show how the meeting of these two technologies allows the change of life in the cities to a life: more sustainable, more egalitarian and more future-oriented. Smart city is a new concept of urbanization that develops rapidly, and it offers novel solutions to the problems triggered by the rapid growth of population, scarcity of resources and environmental deterioration. Smart city the essence of a smart city is to strategically apply and voluntary develop technology and data-infused urban environments aimed to improve quality of life, efficiency of urban operation and services, and competitiveness, while ensuring sustainable growth. The chapter also elaborates on the approaches, technologies and innovations that support smart city administration and their importance in achieving SDGs (Sustainable Development Goals) [16].

1.6.4 The Pillars of Smart Cities

Smart cities are based on series of core pillars which leads to efficient urban management and sustainability. These include:

1. Smart Government: Transparent, and participative decision-making processes through e-government Applications and electronic platforms.
2. Cyber-Physical Systems: The embedding of pervasive and networked information and communication technology into urban infrastructure to increase efficiency, lower resource consumption, and enhance the quality of life.
3. Smart Economy: Promoting innovation, entrepreneurship and digital market connectedness to stimulate economic growth [17].
4. Smart Mobility: Adoption of intelligent transportation systems (ITS), electric vehicles (EVs), and shared mobility services to minimize congestion and carbon emissions [2].
5. Smart Environment: Use of green technology and techniques to reduce environmental impact and boost resource efficiency.
6. Smart Living: Enhancing urban livability via access to excellent healthcare, education, and housing.

1.6.5 Innovations in Smart City Management

Smart cities are an amalgamation of technology with innovative ideas for administration. The following are some key paradigm shifts for sustainable urban development:

- IoT (Internet of Things): IoT equipment connect city systems and provide real-time data for efficient resource management. It includes smart meters, environmental sensors and networked traffic system for instance.
- AI/ML: AI driven analytics for predictive maintenance, efficient utility use and smart urban planning.

- Big Data Analytics: By analyzing large databases, municipal administrators may spot patterns, forecast trends, and make educated choices.
- Blockchain Technology: Blockchain provides data security and transparency in sectors including digital identity management, urban supply chains, and e-governance.
- Geospatial Technology: GIS and remote sensing technologies serve in urban planning, disaster management, and infrastructure construction.
- Digital Twins: Virtual models of cities imitate real-world settings to improve operations and prepare for future growth.

1.6.6 Sustainable Development via Smart Cities

Smart cities play a crucial role in attaining sustainable development by addressing social, economic, and environmental concerns. The following sections demonstrate how smart city projects contribute to the SDGs:

1. Affordable and Clean Energy (SDG 7):
   - Implementation of renewable energy options such as solar panels, wind turbines, and microgrids.
   - Smart grids enhance energy delivery and decrease losses.
2. Industry, Innovation, and Infrastructure (SDG 9):
   - Deployment of intelligent infrastructure that responds to changing urban needs.
   - Promotion of startups and innovation clusters to stimulate technical progress.
3. Sustainable Cities and Communities (SDG 11):
   - Development of eco-friendly structures with smart energy systems.
   - Urban regeneration initiatives that incorporate green spaces and pedestrian-friendly architecture.
4. Climate Action (SDG 13):
   - Use of AI and IoT to monitor and manage urban heat islands, air pollution, and greenhouse gas emissions.
   - Adoption of circular economy ideas to decrease waste.

1.6.7 Case Studies in Smart Cities Management

Numerous cities across the globe are used as benchmarks for smart city projects. The following examples illustrate how technology and managerial strategies are transforming cities:

1. Barcelona, Spain: Barcelona uses IoT-based sensors to manage parking, waste collection and street lighting. The city's intelligent water management system reduces water by 25%.

2. Singapore: Singapore uses digital twinning to simulate urban systems and enhance planning through its Smart Nation initiative. The city also uses AI in health care to improve public health care.
3. Amsterdam, Netherlands: The "Smart City" project in Amsterdam is centered on innovation through cooperation in, for instance, energy, sustainability (transport and houses). They are in fact experimenting with tasting programs.
4. San Diego, USA: The city uses smart lighting which has cameras and sensors to track traffic and air quality, reducing energy consumption by 60%.

1.6.8 Challenges in Smart Cities Management

Smart cities, although promising, face many challenging problems that need to be tackled in a strategic way:

- **Privacy and Security of Data**: Security and data protection are important aspects with increasing number of IoT devices and cyber threats nowadays.
- **Digital Divide**: Fair and equal access to digital infrastructure and services is still an issue in many areas.
- **High Upfront Costs**: Adoption of smart technology requires high capital investment, which might be a barrier for underdeveloped as well as developing communities.
- **Interoperability Problems**: Connecting together different systems and technologies, can be difficult and costly.
- **Responsible Use of Technology**: The e-waste and energy usage linked to digital technology should be used thoughtfully.

Figure 1.4 illustrates the percentage improvement across key urban challenges. Waste Management and Energy Inefficiency show the highest levels of improvement at 27% and 23%, respectively, highlighting successful intervention areas. Traffic Congestion and Water Scarcity show relatively lower improvements, indicating room for further development.

1.6.9 Future Directions in Smart Cities Management

Trends and breakthroughs will dictate the evolution of smart cities. Areas of focus include the following:

1. AI-Powered Urban Analytics: Sophisticated AI models will provide more in-depth understanding of urban issues, enabling to tackle issues proactively.

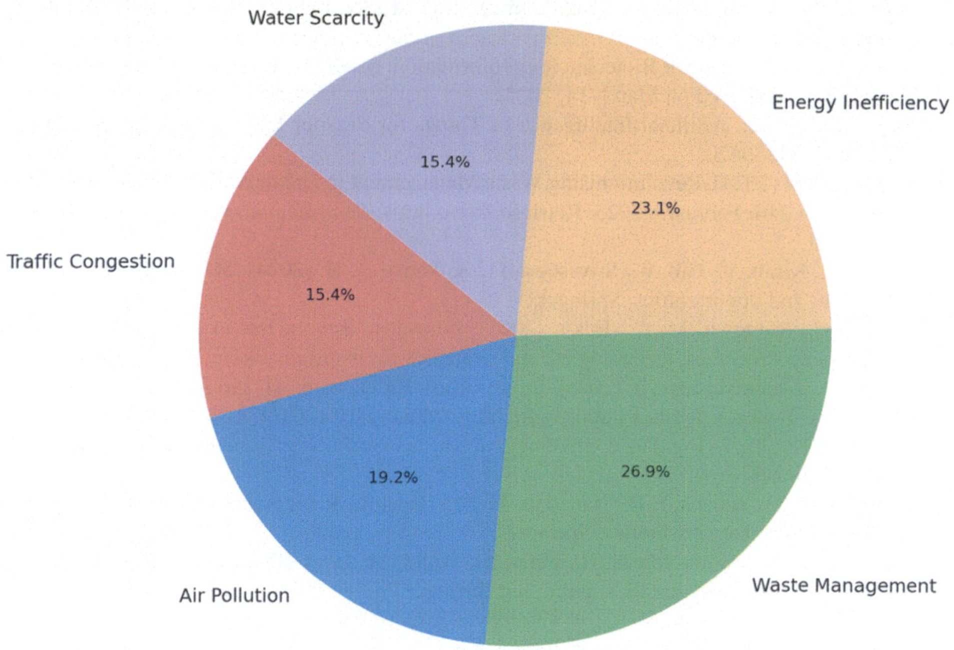

Fig. 1.4 Urban challenges versus improvement

2. Integration in the Circular Economy: Cities will apply the principles of the circular economy to increase resource productivity and reduce waste.
3. Citizen-Centred Design Lean over Citizen design, and focus on citizens collaborating with place and service outcomes through participatory UX platforms.
4. Infrastructure resilience: Construction of infrastructure that is resilient to climate impacts, will withstand adverse weather events.
5. International Cooperation: Through international cooperation, exchange knowledge and work toward standards on smart city development.

References

1. MMCA Alumni. (n.d.). Retrieved March 3, 2025, from https://mmcalumni.ca.
2. Abraham A., Prasad S., Alhammadi A., Lestable T., Chaabane F., (2025). Internet of Vehicles and Computer Vision Solutions for Smart City Transformations. Springer Publisher.
3. aiadmin, (2024). Unlocking the Potential of Artificial Intelligence of Things – Exploring the Intersection of AI and IoT for a Smarter Future. Retrieved from: https://aiforsocialgood.ca/blog/unlocking-the-potential-of-artificial-intelligence-of-things-exploring-the-intersection-of-ai-and-iot-for-a-smarter-future. Assessed on 25th March, 2025.

4. Singh, P. D., & Angurala, M. (2024). Integration of cloud computing and IoT: Trends and challenges. IGI Global.
5. NCBI–National Center for Biotechnology Information. (n.d.). Accessed from https://www.ncbi.nlm.nih.gov. Retrieved on March 14, 2025.
6. Bibri, S.E. (2024). Artificial Intelligence of Things for Smarter Eco-Cities. https://doi.org/10.1201/9781003536420.
7. Faster Capital (2024). Revolutionising Waste Management in Urban Areas with M2M Systems. Assessed on 24th February, 2025. Retrived from: https://fastercapital.com/keyword/collection-routes.html.
8. Hooda, S., Kiran, V., Gill, R., Srivastava, D., & Yousif, J. H. (2024). 5G-enabled smart cities: Challenges and opportunities. Springer.
9. Izah, S. C., & Ogwu, M. C. (Eds.). (2025). Innovative approaches in environmental health management: Processes, technologies, and strategies for a sustainable future. Springer.
10. Sharma B., Gupta M., Jeon G. (2024). Smart Cities; Blockchain, AI, and Advanced Computing. CRC Press. Taylor & Francis Publication. https://doi.org/10.1201/9781003442660.
11. Pandey, B. K. (2024). Big data computing: Advances in technologies, methodologies, and applications. Springer.
12. Reddy, K. K. C., Anisha, P. R., Hanafiah, M. M., Doss, S., & Lipponen, K. (2024). Sustainable energy solutions for smart cities. Springer.
13. Wajid, M. S., Terashima-Marín, H., Zafar, A., Wajid, M. A., & Bhatti, M. A. (2024). Machine learning and data analytics for smart cities. Springer.
14. Nwokolo, S. C., Obiwulu, A. U., & Okonkwo, P. C. (2024). Emerging technologies for smart cities: Innovations and applications. IGI Global.
15. Mutambara, A. G. O. (2023). Artificial intelligence: A driver of inclusive development. Springer.
16. Mhlanga, D., & Dzingirai, M. (2024). Financial inclusion and sustainable development: A global perspective. Springer.
17. Verified Market Research. (n.d.). Retrieved February 10, 2025, from https://www.verifiedmarketresearch.com.

Green Internet of Things (G-IoT) for Sustainable Cities of the Future

2

2.1 Overview of IoT and Sustainable Cities

In an increasingly urbanized world, cities face mounting challenges in balancing growth with sustainability. Internet of Things (IoT) refers to the network of physical objects—sensors, devices, vehicles, appliances—embedded with electronics and connectivity that enable them to collect and exchange data in real time.

These "smart" devices can monitor conditions, communicate over the internet, and be remotely controlled or automated. Sustainable cities (often called smart sustainable cities) are urban environments designed to meet present needs without compromising the ability of future generations to meet theirs. This involves achieving a harmonious integration of environmental protection, social well-being, and economic development [1].

In practice, a sustainable city pursues reduced pollution and resource consumption, improved quality of life for all citizens, and robust, inclusive economic growth.

Bringing IoT and sustainability together yields the concept of the Green Internet of Things (G-IoT)—IoT systems and applications explicitly geared toward sustainability goals. G-IoT encompasses both greening of IoT (making IoT devices and networks energy-efficient and eco-friendly) and greening by IoT (using IoT to help achieve sustainability targets in various sectors) [2]. In sustainable cities, IoT serves as a critical infrastructure layer that can collect data on urban conditions, optimize operations, and enable data-driven decision-making for sustainability [3].

For example, a city deploying thousands of IoT sensors can continuously monitor energy usage, traffic flow, air and water quality, waste levels, and other parameters. The data from these sensors allows city managers and systems to anticipate problems and respond proactively—preventing waste, reducing pollution, and improving services.

V. P. Gupta et al., *Green IoT and AI for Sustainable Development of Smart Cities*,
Synthesis Lectures on Sustainable Development,
https://doi.org/10.1007/978-3-031-98743-4_2

2.2 Relevance of IoT to Sustainable Cities

IoT technologies support all three pillars of sustainability:

- **Environmental sustainability**: IoT is widely used in environmental monitoring and resource management. By detecting changes in conditions such as air quality, noise, temperature, or water pressure, IoT systems help identify pollution spikes, leaks, or other inefficiencies in real time.

  For instance, as per the CCAC Secretariat (hosted by the UN Environment Programme). London installed 100 IoT sensors in 2019 to monitor air quality across the city. Similarly, Helsinki's university deployed an IoT-based system to track airborne particulate matter. Continuous monitoring means that cities can respond faster to environmental issues—issuing smog alerts, fixing water leaks, or adjusting traffic flows to cut emissions. IoT-driven management of energy and resources also reduces wastage. A smart electricity grid in Helsinki, Finland, resulted in a 15% reduction in energy consumption through better demand-response and efficiency measures (UNEP. IPCC Reports. 1990, 1995, 2001, 2007) [4]

  Collectively, such improvements curb greenhouse gas emissions and help cities address climate change. (In fact, cities account for over 75% of global CO_2 emissions, so enhancing urban efficiency via IoT can have a significant climate impact.) Furthermore, IoT sensors strengthen disaster resilience by providing early warning for natural hazards. Many cities now deploy networks of environmental sensors that detect earthquakes, floods, or wildfires and automatically alert authorities and citizens, greatly improving response times. In 2020, 23 UN member countries had effective IoT-powered early warning systems that protected over 93% of their at-risk populations from natural disasters [5]. Social sustainability: A sustainable city is also one that fosters a high quality of life, equity, safety, and inclusive services for its residents. IoT contributes to social sustainability by enabling smart community services, transparent governance, and improved public safety. By publishing open data from IoT sensors (for example, live air quality or transit data on city dashboards), city governments can engage citizens with information and build public trust. When residents have access to real-time information and see improvements (like shorter commutes or cleaner air) resulting from smart initiatives, they become more confident in city management and more involved in civic life. IoT also plays a direct role in public safety and health. Cities like Singapore have installed connected cameras and sensors in public areas, which, when combined with analytics, have dramatically reduced crime rates—Singapore famously reported zero robberies or thefts in a 322-day period after implementing an IoT-enhanced surveillance program. In healthcare, wearable IoT devices and remote patient monitors help doctors keep track of vulnerable patients' vital signs and conditions. Such smart healthcare applications enable early detection of medical issues and improve emergency response. For example, IoT patient monitors in hospitals can alert staff to subtle signs of deterioration so that interventions occur before a crisis. IoT connectivity is also valuable in disaster and emergency response from a social perspective: smart devices and networks can rapidly disseminate warnings to the public and coordinate first responders. After an incident, IoT devices (like smartphones or wearables) can help locate survivors by detecting mobile signals or vital signs, significantly improving search-and-rescue operations within the critical 72 h "golden window" [6].

- **Economic sustainability**: Economic vitality in a city is enhanced by IoT through greater efficiency, innovation, and cost savings. Smart city systems reduce operational costs for municipalities (for instance, optimizing energy use in street lighting or water distribution saves money) and they often deliver services more efficiently, which benefits businesses and residents. IoT-based smart grids, for example, minimize energy loss and downtime—ensuring businesses have reliable power and consumers pay only for what they use.

  Smart meters can enable dynamic pricing that flattens peak demand and lowers utility bills. In transportation, IoT reduces the economic loss of time spent in traffic by easing congestion. According to one estimate, an AI-driven traffic management system in Kansas City (USA) that leverages IoT sensors to adjust traffic lights cut average travel times by up to 25%, allowing commuters to save time and boosting productivity. IoT-driven optimization also means infrastructure is used more effectively, potentially deferring the need for costly new construction. Additionally, the push for G-IoT creates economic opportunities in the technology sector: developing and deploying sensors, networks, and data analytics creates green jobs and spurs innovation. By improving resource efficiency (doing more with less), IoT helps cities achieve economic gains while using fewer resources—a core principle of sustainability [7].

  So, IoT serves as a nervous system for smart sustainable cities, continually sensing the urban environment and enabling adaptive, efficient management.

2.3 Emerging Trends and Technologies in G-IoT for Smart Cities

Building a sustainable smart city requires not only applying IoT to new use cases but also innovating the IoT technology to be energy-efficient, resilient, and integrated with advanced capabilities.

In recent years, several key trends and technologies have emerged in what can be called "Green IoT" for smart cities:

- **Energy-Efficient IoT Devices**: Designing IoT hardware with **low-**power electronics and protocols is a cornerstone of G-IoT. Many urban IoT sensors need to operate for years in the field (often on battery power) [8]. Advances in microelectronics

Table 2.1 Key differences between traditional approach (Pre-IoT) and G-IoT-enabled approach

City function	Traditional approach (Pre-IoT)	G-IoT-enabled approach
Electric power	One-way power flow; manual meter readings; periodic outage checks	Two-way smart grid with sensors and smart meters enabling real-time balancing of supply/demand and quick fault detection. Consumers can feed in solar power; utilities adjust to usage patterns dynamically
Water management	Water usage measured infrequently; leaks often go undetected until major issues; fixed pump schedules	Smart water system with IoT meters and leak sensors providing continuous usage data and instant leak alerts. Real-time monitoring of water quality and pressure helps prevent losses (addressing the ~ 25–30% water lost as "non-revenue water" in many cities)
Transportation	Static traffic signal timings; limited data on congestion; reactive response to accidents	Smart traffic systems using IoT sensors and cameras to monitor flow and incidents in real time. Traffic signals adjust adaptively to reduce congestion (e.g., AI-controlled lights improved travel times by 25% in Kansas City). Integrated apps guide drivers to free parking, cutting unnecessary circling
Street lighting	Lights on fixed timers or always on at night, regardless of actual need	Smart lighting with IoT-connected LED streetlights that dim or brighten based on ambient light or motion sensors. This cuts energy use dramatically and allows remote outage detection
Waste collection	Scheduled pickups regardless of bin fill levels; routes set manually; overflowing bins between pickups	Smart waste management with sensor-equipped bins that report fill status. Collection routes and frequency are optimized based on need, reducing fuel use and overflow incidents. (Studies show IoT waste systems can cut collection costs ~ 50% and carbon emissions ~ 20%)
Community alerts	Reliance on broadcast alerts (TV/radio) or sirens for emergencies; one-way communication	Connected alert systems that push notifications to smartphone apps, digital signage, and IoT sirens. Citizens receive location-specific alerts (e.g., for severe weather or hazards) and can also send feedback or SOS signals via connected devices

Source Created by Authors

and firmware allow modern sensors and communication modules to consume only micro-watts or milli-watts of power in standby, and to transmit data efficiently. For examples, Next-gen BLE beacons and Zigbee/Thread devices in smart buildings remain in sleep mode for the most part, only waking up to transmit data, thereby significantly increasing battery life. Low-power chipsets and intelligent design enable some sensors to operate 5–10 years on a coin-cell battery. Low-power wide-area networks can similarly accommodate small battery-powered devices for up to 10 years. Long lifespan facilitates sustainability by lowering battery waste and maintenance. Moreover, manufacturers are looking at green materials and recyclable product designs for IoT devices.

- **Solar-Powered and Energy Harvesting Sensors**: G-IoT extends beyond batteries by tapping renewable and ambient sources of energy to activate devices. Solar-powered sensors, for instance, are possible to use continuously without grid power, which suits smart city applications such as traffic or environmental sensing. Various other sources of energy like heat, motion, RF signals, and wind are also being utilized to activate small IoT devices. A common example is a solar-powered air quality sensor that recharges by day and transmits data without battery replacement. Such energy-harvesting techniques facilitate cleaner IoT systems by reducing battery waste and grid consumption. Efforts towards ongoing standardization focus on enhancing energy storage and compatibility, making the way for increasingly self-sufficient outdoor sensor networks for weather, pollution, and wildlife tracking.
- **AI Integration (AIoT)**: A major trend is the convergence of IoT with artificial intelligence—often termed **AIoT**. In sustainable smart cities, AI and IoT work together in symbiosis: IoT provides the data streams, and AI provides the analytics and decision-making. Embedding AI into IoT systems allows cities to not only collect data, but interpret it and respond autonomously. An intelligent building HVAC system utilizes IoT sensors to gather data on temperature, occupancy, and air quality, and AI to set comfort and energy-saving settings. At a city level, AIoT provides real-time traffic management, forecasts infrastructure maintenance requirements, and predicts energy or water consumption. AI also optimizes IoT network efficiency by streamlining data routing and compression. This integration provides systems with increased adaptability and efficiency, allowing cities to efficiently use resources and enhance services such as traffic movement and public safety.
- **Low-Power Wide-Area Networks (LPWAN)**: To connect the multitude of distributed sensors across a city in an energy-efficient way, LPWAN technologies have become indispensable. LPWANs (such as LoRaWAN, Sigfox, NB-IoT, and LTE-M) are communication networks optimized for low-bandwidth IoT devices that need to transmit small amounts of data over long distances while consuming minimal power [9]. Their characteristics—long range, penetration through buildings, and multi-year battery life—perfectly suit city-scale deployments like smart streetlights, water meters,

parking sensors, environmental sensors, and more. For example, LoRaWAN and Sigfox networks can often reach 5–15 km ranges in urban areas and support thousands of devices on a single gateway, all while devices only sip microamps of power when idle. NB-IoT (Narrowband IoT) is a cellular LPWAN standard that operates on licensed bands and can similarly support massive numbers of devices with very low power consumption, even in underground or indoor locations.

The benefit of LPWAN for sustainability is two-fold:

a. IoT devices can operate untethered (on small batteries or harvesters) for long periods, which reduces maintenance and battery waste.
b. LPWANs are power-efficient and cover large distances with low base stations, making them suitable for smart cities. They offer a stable backbone for connecting devices such as utility meters, air quality sensors, and agriculture monitors. Serving as the city's "sensory organs," LPWANs facilitate data flow from remote sensors while saving power, enabling large-scale, environmentally friendly IoT deployments.

5G Connectivity and Future Opportunities in Smart Cities
In addition to LPWAN for IoT devices with low power, 5G presents new opportunities for high-speed and secure communication in cities. It enables three primary types of communications—enhanced broadband, ultra-reliable low-latency links, and massive machine connections. The latter pair are particularly priceless in intelligent cities, where intelligent transportation systems like autonomous cars, drones, and factory controls utilize real-time exchange of data and respond within fractions of a second, as fast as a few milliseconds. Both vehicle-to-infrastructure communication and fleet management will hugely benefit from such low-latency support.

5G is also unique in that it can support millions of devices in a limited space, which is crucial as cities expand their IoT infrastructure. With capabilities such as network slicing, specific channels can be reserved for mission-critical services like emergency response, providing guaranteed performance even during heavy traffic. This flexibility ensures that service quality is preserved while it accommodates an expanding network of sensors and connected devices.

Although 5G is not as power-efficient as LPWAN, cities are heading towards a hybrid solution. High-data or time-sensitive devices such as video surveillance systems or intelligent vehicles use 5G, whereas low-complexity, battery-powered sensors continue on LPWAN networks. This solution offers an optimal, cost-effective solution to service varied connectivity demands throughout the city.

Collectively, technologies such as LPWAN, 5G, and forthcoming advancements like 6G and satellite IoT are building a robust foundation for smart city infrastructure. Coupled with energy-efficient devices and AI-driven analytics, they make urban IoT systems not just smarter and quicker but also sustainable and responsive to real-time requirements.

Example: A city park equipped with modern IoT shows how emerging tech supports sustainability. Solar-powered soil sensors send data via LoRaWAN, while AI predicts irrigation needs and smart valves use only the required water. A 5G drone monitors tree health and cleanliness, helping target maintenance efficiently. Together, these tools demonstrate how G-IoT can optimize resources and city operations.

2.4 Smart Development: G-IoT in Infrastructure and Urban Planning

Smart cities weave G-IoT into urban planning and infrastructure as an overarching strategy, not merely as independent projects. Infusing IoT into roads, buildings, utilities, and public areas enables continuous data gathering. Such real-time information enables wiser planning decisions, enabling cities to become more efficient, waste-reducing, and more liveable.

This section shows how Green IoT enables smarter infrastructure and planning through data-driven, sustainable decisions:

2.4.1 Smart Traffic and Transportation Systems

Urban traffic congestion leads to environmental harm, economic loss, and commuter frustration. Green IoT helps cities address this by enabling smart traffic management. Sensors like cameras, GPS devices, and road loops provide real-time data to traffic systems, which use it to adjust signals, manage routes, and support public transport efficiency. Kansas City demonstrates this well—its AIoT-based traffic grid analyzes live conditions and adapts signal timing, reducing travel times by up to 25%. This results in less idling, fewer emissions, and faster emergency responses. Additionally, smart parking systems use sensors or cameras to detect available spaces, guiding drivers via apps and cutting unnecessary driving. Public transportation also benefits, with many cities using IoT trackers on buses and trains to provide real-time arrival info on digital displays and mobile apps. Altogether, IoT makes urban transport systems more adaptive, sustainable, and user-friendly compared to traditional, fixed-schedule approaches.

2.4.2 Green Buildings and Smart Buildings

Urban buildings contribute significantly to a city's energy usage—typically between 30 and 40%. Due to this, converting conventional buildings into intelligent, energy-saving spaces is becoming increasingly important in sustainable urban planning. One of the drivers for this change is the Internet of Things (IoT), which upgrades Building Management Systems (BMS) with the addition of smart sensors and automation to optimally manage lighting, heating, cooling, and air systems.

For instance, in commercial areas, sensors may be used to sense occupancy and modulate the HVAC systems to cool or heat occupied rooms only. This targeted control prevents waste of energy and ensures comfort. Research has indicated that these smart HVAC systems can save an estimated 20–25% electricity consumption. Likewise, IoT-led lighting based on motion and daylight sensors can reduce lighting energy consumption by as much as 40% by varying brightness or turning off lights in vacant spaces.

Cities are increasingly implementing retrofitted solutions such as smart thermostats, automated window shades, and air quality sensors in residential and commercial buildings. These technologies not only reduce energy expenses and emissions but also enhance indoor comfort and building safety. They also allow for predictive maintenance by identifying structural problems early, encouraging long-term resilience and sustainability.

2.4.3 Environmental Monitoring and Urban Green Spaces

Proper urban planning these days is entirely dependent on detailed, real-time environmental data. Due to the Internet of Things (IoT), cities can use widespread networks of sensors monitoring air quality measurements such as PM2.5, nitrogen dioxide, and ozone, in addition to noise, temperature, humidity, and even soil and wildlife in parks. In contrast to older systems with few monitoring stations, IoT offers high-density data coverage at many locations, providing planners with a more complete and detailed environmental picture. This ongoing stream of data allows for focused interventions. For example, endemic air pollution in particular locations may necessitate traffic adjustments or the introduction of low-emission zones. Noise levels can initiate measures like the imposition of quiet periods or the redesign of roads to cut back on sound pollution.

Urban green spaces also enjoy the advantages of smart technologies. Sensors for soil moisture assist in irrigation management to conserve water. At the same time, foot traffic information can help design improved park spaces or inform future expansions. Furthermore, several cities are embracing digital twins—dynamic virtual replicas fueled by real-time IoT data—to simulate development plans before execution, building smarter, greener, and more resilient cities.

2.4.4 Data-Informed Urban Planning

Modern urban planning relies heavily on precise, real-time environmental data. With the integration of IoT, cities now deploy extensive sensor networks that monitor air pollutants like PM2.5, nitrogen dioxide, and ozone, along with ambient noise, temperature, humidity, and even ecological factors in parks. Unlike traditional methods with limited data points, IoT offers widespread, detailed insights across urban spaces.

This data stream helps identify pollution-prone zones and supports targeted responses—such as altering traffic flow or creating low-emission areas. Elevated noise levels might prompt quiet zone policies or road redesigns to reduce disturbances.

Green spaces benefit as well. Soil sensors optimize irrigation, conserving water while keeping landscapes healthy. Data on park usage guides better facility planning and green space expansion. Additionally, cities increasingly use digital twins—virtual city models powered by IoT data—to test development scenarios and support sustainable, informed decision-making.

In short, by embedding IoT into infrastructure (making it "smart") and by employing IoT data in urban planning, cities create a virtuous cycle: *smart infrastructure generates data → data drives better planning → better planning yields more sustainable infrastructure and services.*

Table 2.2 gives a snapshot of the use of Green IoT (G-IoT) in different sectors of urban infrastructure, along with the most significant advantages these technologies have to provide. It showcases how smart solutions help bring enhanced efficiency, sustainability, and quality of life to cities.

By embracing these technologies, cities are undertaking "smart development"—development that converges innovation with sustainability. No longer do city planners view IoT as a mechanism, but rather as a given infrastructure. Like the saying, "data is the new asphalt": just like roads and pipelines defined cities throughout the previous century, sensor systems and real-time information are defining sustainable cities today.

2.5 G-IoT in Water and Energy Conservation Management

Two of the most critical resource systems in any city are water and energy. Managing these systems sustainably is essential for a city's future, and G-IoT provides powerful tools to conserve resources, improve efficiency, and ensure reliability [10].

So, we delve into how Green IoT is revolutionizing water management (including water supply, wastewater, and irrigation) and energy management (particularly electricity via smart grids and smart meters) (Fig. 2.1).

2.5.1 Smart Energy Grids

The traditional electrical grid is undergoing a transformation into a **smart grid**—essentially an IoT-enabled power network. A smart grid uses digital communications and IoT sensors throughout the power system: from generation plants and renewable energy farms,

Table 2.2 Examples of G-IoT applications across urban infrastructure sectors

Sector	G-IoT Application Example	Sustainability Benefits
Transportation 	Smart traffic lights and IoT-enabled congestion management; smart parking systems; connected public transit (buses with GPS)	Reduced traffic jams and vehicle idling = lower emissions; shorter commute times improve economic productivity and quality of life; public transit reliability encourages its use (cutting private car usage)
Buildings 	Smart building management systems (HVAC and lighting automation via IoT sensors); occupancy-based climate control; IoT-enabled energy management platforms	Significant energy savings (10–20% less building energy use overall); lower carbon footprint; improved indoor comfort; data on energy use helps target retrofits and renewable integration (e.g., solar)
Energy grid 	Smart grid with IoT sensors along power lines and substations; smart electricity meters at homes; demand-response IoT devices (smart thermostats, EV chargers)	Higher grid efficiency and reliability (faster outage detection and restoration); integration of renewables (IoT helps balance variable solar/wind input) consumers are more aware of and can adjust their energy usage, saving costs and energy
Water and wastewater 	IoT water meters and pressure sensors; smart leak detection in pipelines; IoT irrigation controllers; wastewater treatment sensors	Water conservation through quick leak repair and optimized usage (addressing water scarcity); lower energy usage in water distribution; safer water quality through continuous monitoring; optimized irrigation avoids over-watering city landscapes
Waste management 	Sensor-equipped waste bins and dumpsters; route optimization software; AI-based sorting robots in recycling facilities; fill level alerts to waste collectors	Fewer collection trips (fuel and emission reduction—e.g., one city saw a 20% CO_2 cut with smart waste systems); less overflowing trash (improving hygiene); increased recycling rates due to better sorting and monitoring (30–50% increase reported with smart systems)

(continued)

Table 2.2 (continued)

Sector	G-IoT Application Example	Sustainability Benefits
Urban planning 	City digital twin fed by IoT data; open data dashboards showing traffic, pollution, energy usage; crowd-sourced reporting apps linked with IoT (e.g., noise reporting)	More informed decision-making for sustainable development (data-driven); ability to simulate impacts of projects and choose greener options; improved public engagement and transparency builds support for sustainability initiatives
Public spaces 	Environmental sensors in parks (air quality, UV index); smart lighting and irrigation in public areas; pedestrian flow sensors on sidewalks	Healthier environments (e.g., identifying polluted areas to mitigate); reduced resource use (water, electricity) in maintaining public spaces; data to design more accessible, pleasant urban spaces (e.g., ensure adequate lighting or shade where people need it)

through transmission lines and substations, down to distribution in neighbourhoods and even within homes (smart meters and smart appliances). This results in a two-way flow of both electricity and information.

For instance, smart meters placed in residential homes and commercial establishments constantly monitor electricity consumption and report back to utilities at regular intervals—often every few minutes. These finer-grained data enable utilities to study consumption patterns, spot problems such as outages or equipment malfunctions in real time, and react quicker. It also enables demand-response measures, wherein smart devices can briefly lower power consumption during peak demand (e.g., temporarily shutting off an HVAC system) to alleviate pressure on the grid and avoid blackouts.

- From a sustainability perspective, this IoT-powered energy management provides various important advantages:
- Enhanced grid efficiency, through more accurate balancing of supply and demand.
- Less energy waste, through real-time monitoring and remote control.
- Reduced carbon emissions, through the incorporation of additional renewable energy sources and reducing dependence on fossil fuel-powered peaker plants.
- Consumer empowerment, as customers receive information about their energy usage and are able to make more educated, environmentally friendly decisions.

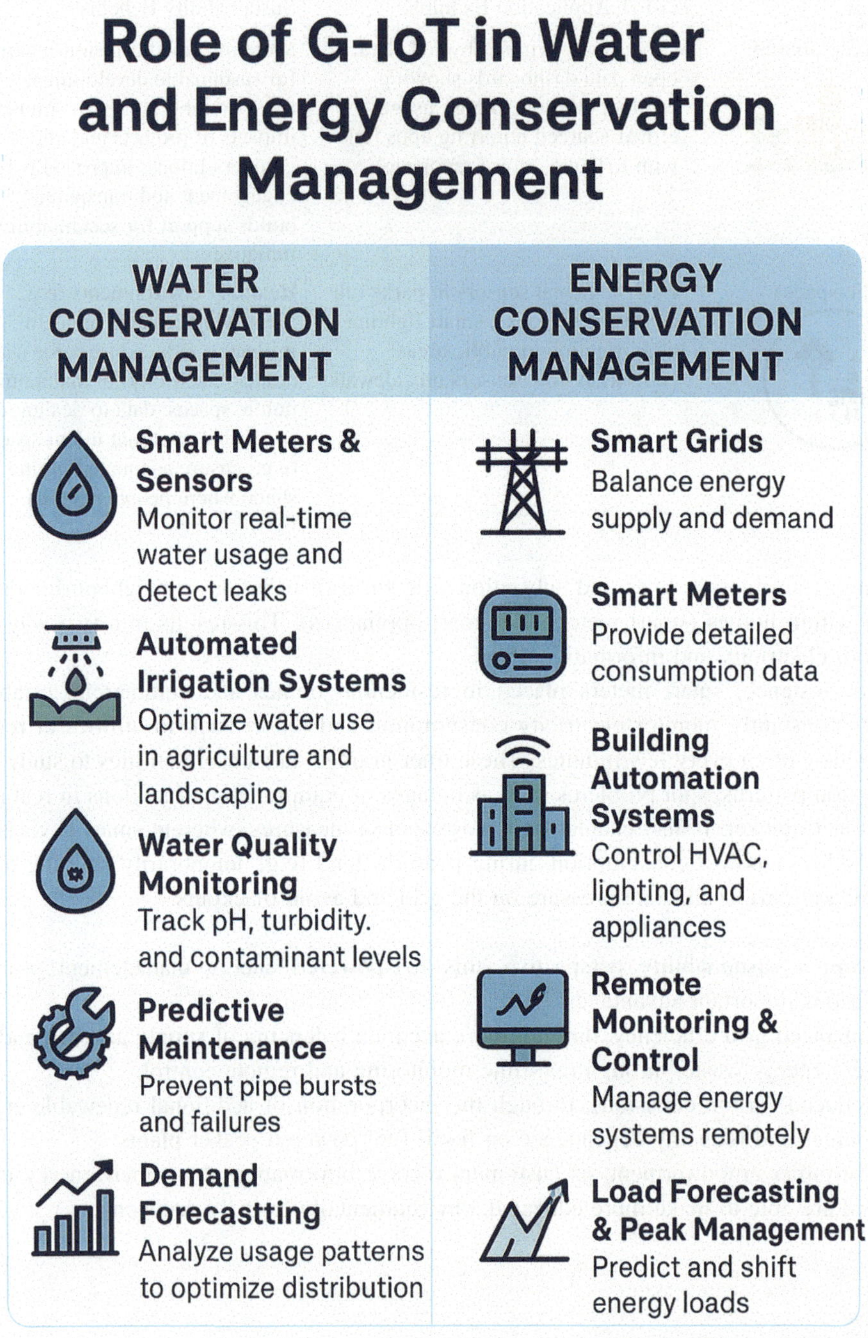

Fig. 2.1 Water conservation management v/s energy conservation management [11]

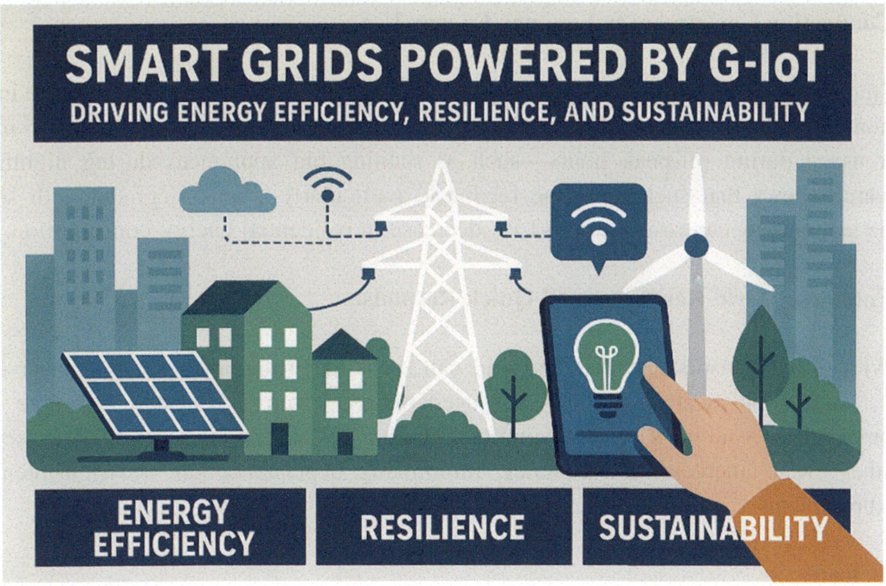

Source Author's Creation

2.5.2 Smart Grids Powered by G-IoT: Driving Energy Efficiency, Resilience, and Sustainability

- **Improving Energy Efficiency and Minimizing Losses**.

Grid-IoT (G-IoT) infrastructure ensures constant monitoring of energy grids, which simplifies the detection of problems such as system inefficiencies, overheated components, or misuse. Automated systems enable precise adjustment of voltage and frequency, minimizing wasteful energy loss. A smart grid program in Helsinki, to cite an example, resulted in a 15% reduction in energy consumption—demonstrating the potent effect of these technologies [12].

Facilitating Easy Integration of Renewable Energy
Sources such as solar and wind are inherently variable, but G-IoT technologies mitigate this by dynamically balancing supply. If solar output is low due to cloud cover, the system can instantly switch to stored energy or flexible generators to ensure grid stability. This flexibility enables a higher proportion of renewables in the energy mix, allowing cities to reduce emissions and move towards cleaner objectives.

- **Connecting Consumers to Smarter Energy Use**.

Smart meters and mobile applications provide consumers with minute-by-minute information about their electricity usage. Time-of-day pricing encourages individuals to move their usage during off-peak hours—such as running big appliances during nighttime. Research shows that such real-time feedback, particularly when integrated with smart home technology, can result in a 5–15% decrease in residential energy consumption.

- **Enhancing Grid Resilience and Quick Response**.

With G-IoT sensors installed, the grid can sense and respond to faults in near real-time. Some sophisticated networks can even redirect electricity around a trouble spot automatically, avoiding service disruptions. This quick response is essential as severe weather events increase in frequency, minimizing dependence on emission-intensive backup systems.

Conclusion
G-IoT smart grids lay the foundation for productive, sustainable, and secure urban energy networks. Through optimization, integration of renewable energy, and engagement with customers, they create the pillars of a low-carbon, future-proofed infrastructure.

2.5.3 Water Conservation and Smart Water Management

Water is another resource under pressure (due to growing demand and climate-related scarcity). IoT-based **smart water management** is emerging as a solution for cities to monitor and conserve water comprehensively.

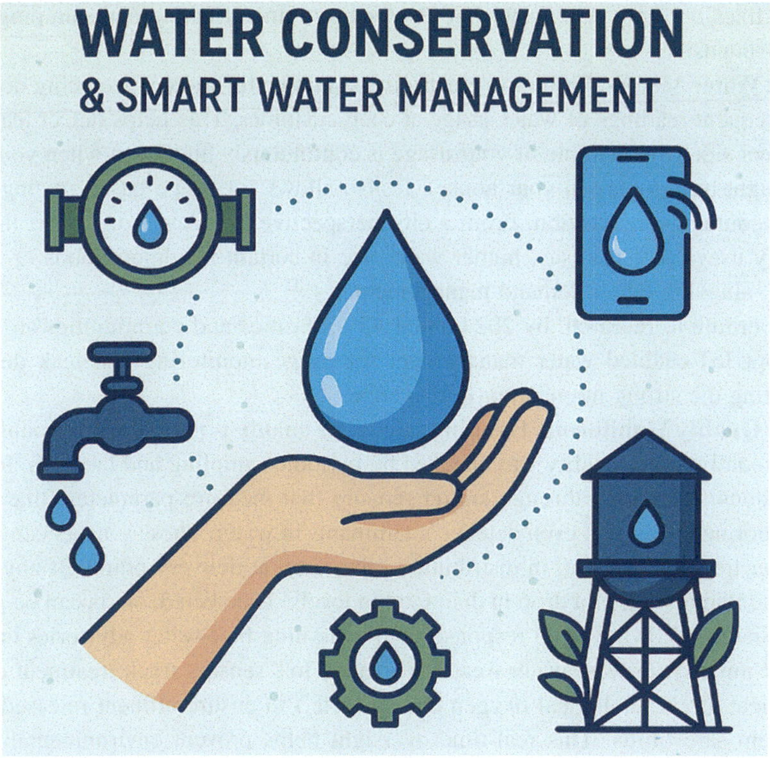

Source Author's Creation

Key aspects includes:

- Leak Detection and Loss Reduction: A startling amount of urban water is lost before it ever reaches consumers—this "non-revenue water" is due to leaks in pipes and theft or meter inaccuracies. The World Bank estimates that globally about 25–30% of water is lost in distribution networks. IoT helps tackle this by installing pressure sensors and flow meters throughout the network. By analyzing the data, the system can pinpoint unusual drops in pressure or discrepancies in flow that indicate a leak, often narrowing it down to a specific pipe segment. Smart water platforms send instant leak alerts to utility teams, who can dispatch repair crews before a major pipe burst or significant water waste occurs. In smart city pilots, such IoT leak detection systems have saved millions of gallons of water and reduced pipe break incidents by enabling preventive maintenance [13]

  For instance, one IoT solution might notice that water flow at midnight in a district is higher than expected (when usage should be minimal) and flag a probable leak.

Quick fixes not only save water but also energy (since treating and pumping water is energy-intensive).

- **Smart Water Meters**: Similar to electricity, **smart water meters** are being deployed to give frequent readings of water usage at each customer. This helps detect leaks on the customer side (for example, if your usage is continuously high even when you're away, you might have a leak in your house). It also allows for usage-based pricing schemes that encourage conservation. From a city perspective, the data from smart meters can identify usage patterns (say, higher water use in certain neighbourhoods or at certain hours), allowing better demand management.

  According to research, by 2024 about 45% of cities and communities are expected to adopt IoT-enabled water management for usage monitoring and leak detection—indicating the strong momentum in this area.
- **Water Quality Monitoring**: Ensuring safe water quality is paramount for public health. Traditionally, water quality was checked by periodic sampling and lab tests. IoT brings continuous monitoring through **smart sensors** that measure parameters like turbidity, pH, chlorine levels, and even detect contaminants in water. These sensors can be placed at water treatment output, in distribution pipes, and at delivery points. If any anomaly (e.g., contamination or a drop in disinfection levels) is detected, alerts can be generated immediately, allowing rapid response such as issuing boil-water advisories or flushing lines. Similarly, in wastewater treatment plants, IoT sensors track treatment efficiency (chemical levels, biological oxygen demand, etc.) to ensure effluent released to rivers is within safe limits. This real-time oversight helps prevent environmental pollution events and can protect ecosystems.
- **Precision Irrigation and Landscape Watering**: Within cities, a significant amount of water is used for irrigation of parks, public gardens, golf courses, and other green areas. IoT has introduced precision irrigation techniques, similar to those used in agriculture, to urban landscape management. Soil moisture sensors, combined with weather data and smart controllers, ensure that irrigation systems water plants only when and as much as needed. For example, if rain is forecast or if soil moisture is already adequate, the system can skip a scheduled watering cycle, saving water. Conversely, during heat waves the system might water a bit extra to keep plants alive, but at times when evaporation is lowest (e.g., early morning or evening). The savings from such smart irrigation can be substantial—studies in agricultural settings have shown **water savings of 20–30%** or more without sacrificing crop yield, using IoT-based irrigation control. In a city context, this means healthier urban greenery with less water. It also saves the energy that would have been used to pump that water. Some cities are even using IoT-controlled drip irrigation for street trees and planters, directly delivering water to roots with minimal waste.

Smart Energy Usage in Buildings (Demand Side Management)
While the earlier part on smart grids covers the supply side, it's worth noting how IoT aids demand side energy management in buildings and facilities, as part of conservation. Smart thermostats (like Google Nest or others) learn household patterns and adjust heating/cooling schedules to save energy while maintaining comfort. IoT-connected appliances can schedule themselves (e.g., a smart water heater might heat water when grid electricity is plentiful and cheap, and stay off during peak times). In commercial buildings, IoT systems manage energy by turning off lights and equipment in unoccupied zones, and by responding to utility demand-response signals (for instance, temporarily dimming all lights by 10% to reduce load when needed—barely noticeable to occupants but energy-saving at scale). According to a report by the International Energy Agency, digitalization of building controls could cut building energy use by as much as 10% by 2040. These demand-side IoT applications complement the smart grid and contribute to overall energy conservation in the city. In water and energy domains alike, G-IoT provides fine-grained visibility and control that simply was not possible before. The result is a shift from reactive management (fix problems after they become big) to proactive and even predictive management (addressing issues early, optimizing continuously).

2.6 Building Smart Communities: G-IoT for Social Sustainability and Quality of Life

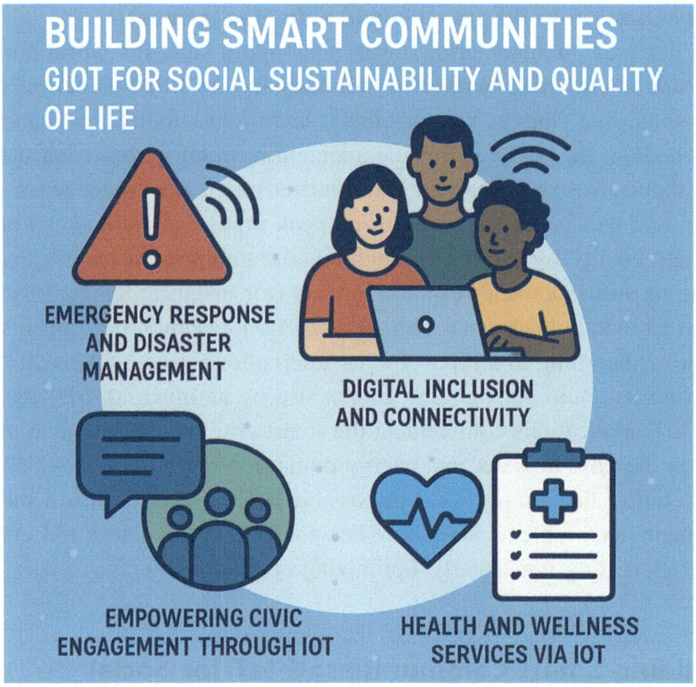

While technology is often discussed in terms of infrastructure and efficiency, its ultimate purpose in a sustainable city is to improve the lives of people. G-IoT, when thoughtfully integrated, can foster smart communities—i.e., communities that are safe, inclusive, well-served by public utilities, and engaged in civic life. IoT is changing city services to be more responsive and people-oriented. In public safety, intelligent systems now utilize connected devices such as CCTV, gunshot sensors, smart streetlights, drones, and emergency wearables. The devices provide real-time information to emergency responders and law enforcement agencies.

Intelligence cameras can sense out-of-normal behaviour or traffic congestion and alert the authorities, and gunshot sensors can detect as well as pin-point shots instantly. IoT-enabled monitoring of infrastructure also enables them to detect damage or threats to bridges and tunnels and enhance both security and safety. IoT hastens emergency response. Internet-enabled cars or roadside units can call in crashes with exact locations and impact, which medics get immediately. Automatic clearing of routes for ambulances can be done through intelligent traffic control. During calamities, flood sensors, quake sensors, and weather sensors provide timely warnings and improve evacuations. IoT also supports community safety. Adaptive streetlights that change in response to pedestrian traffic can deter crime. Air quality monitors can reveal pollution problems in under-resourced neighbourhoods, which can help ensure public health. Even waste collection service is improved. Sensors reduce the risk of garbage overflows, keeping neighbourhoods cleaner and healthier. By becoming more efficient with public services, cities can refocus savings into neighbourhood initiatives such as schools, parks, and social welfare.

- **Emergency Response and Disaster Management**

G-IoT in disaster management involves early detection, fast communication, and effective coordination. Cities prone to earthquakes are installing IoT accelerometers in buildings and underground that can detect tremors and trigger alarms or automatic shut-offs (for gas lines, trains, etc.) critical seconds before the strongest shaking arrives. In coastal cities, IoT-connected sirens and SMS alert systems (triggered by flood sensors or tsunami buoys) can evacuate populations to safety with precious time to spare. During disasters, maintaining communication is vital—IoT mesh networks can be deployed when cell networks fail, connecting first responders' devices and victims' phones in an ad-hoc manner.

- **Digital Inclusion and Connectivity**

A socially sustainable city is one where all residents have access to information and services, including the marginalized or disabled. IoT can promote digital inclusion in multiple ways. City-wide public Wi-Fi or low-cost IoT connectivity can ensure even those without expensive data plans can connect to the internet in public places—helping bridge the digital divide. Some cities use IoT kiosks that offer free internet access, device charging, and city service information, placed in underserved areas. Accessibility for people with disabilities is also improved via IoT, for instance, smart intersection signals that interact with a visually impaired person's smartphone to give audio cues for safe crossing, or IoT-powered navigation apps that guide wheelchair users via routes that are ramp-accessible. Additionally, the data from IoT can inform city programs to address inequality—e.g., foot traffic sensors might reveal that a certain neighbourhood lacks adequate public transportation at certain times, prompting the city to adjust bus services. Inclusive smart city design means involving communities in the planning of IoT deployments and ensuring that new tech doesn't just benefit the tech-savvy or affluent.

- **Empowering Civic Engagement through IoT**

Smart cities become more robust when citizens assist in crafting local solutions. IoT enables this by establishing new means for citizens to engage with their city. Apps or websites are now used by many where citizens can report issues such as potholes or malfunctioning lights, frequently accompanied by sensor data—for example, a report on a streetlight that is out might include its most recent activity to facilitate repairs.

Cities are also becoming transparent by publishing real-time data through public dashboards that display traffic movement, air or water pollution levels, or water consumption. Such openness encourages citizens to suggest ideas or create their own applications based on open data sets. In Singapore, for instance, open smart city data has given rise to innovations such as accessible route mapping for the disabled [14].

Crowdsourced information is another form of contribution by communities. Citizens can have sensors, like air quality sensors, on their land and infuse richness into the city's sensor web while being able to gain essential environmental information themselves. Both awareness and local ownership are increased through this method.

Public planning is also being improved. Cities are now employing visualizations of IoT data—such as speed heat maps at traffic meetings—to make discussions fact-based. This helps build trust, as citizens realize that not only are they heard but their input is also supported by data the city seriously considers.

- **Health and Wellness Services via IoT**

IoT is increasingly becoming a part of urban health, integrating personal care and public health systems. Wearable ECGs, glucose meters, and intelligent blood pressure cuffs are now allowing patients, particularly the elderly, to monitor chronic conditions at home. Cities are increasingly getting together with health providers to supply these devices and implement warning systems that alert authorities to out-of-character behaviors, such as inactivity from an otherwise active residence. IoT also improves public health by facilitating early outbreak detection via anonymized information from smart thermometers or drug purchases. During the COVID-19 pandemic, numerous cities employed IoT and mobile technology for contact tracing and quarantines monitoring, demonstrating how digital technologies can aid large-scale health responses.

Apart from direct care, IoT brings value to daily life—consider smart fitness paths that lead joggers, play areas that verify equipment safety, or engaging displays in public areas such as libraries and museums. These aspects, though nuanced, enhance wellness and help build a healthy community.

So, Intelligent technology must work for humans, not dominate them. Those cities integrating IoT with openness, learning, and citizen input have improved adoption and trust. People are more apt to support and interact with technology when they comprehend and profit from it.

Table 2.3 consolidates the Contributions and Benefits of Green IoT (G-IoT), positioning its value within the three sustainability pillars: social, environmental, and economic. Although emphasis lies with social effects—enhanced public services, healthcare, security, and civic participation—the table also indicates relevant intersections with environmental and economic aspects, demonstrating G-IoT's comprehensive function in creating more intelligent, sustainable communities.

Table 2.3 Benefits of G-IoT mapped to sustainability pillars

Sustainability Pillar	G-IoT Contributions and Benefits	Examples from Smart Cities
Environmental (quality and resilience)	- Continuous monitoring of air and water quality for healthier communities. Lower pollution and emissions through optimized traffic and utilities. Better waste management and cleaner streets/public spaces. Greater resilience to climate impacts and disasters via early warning systems	Ex: London's air quality IoT network guiding policies to improve air (environmental + health benefit); Singapore's IoT-driven waste system reducing landfill use and emissions; Coastal cities using IoT flood sensors to protect neighborhoods
Social (equity, health, safety, inclusion)	- Improved public safety and reduced crime through smart surveillance and lighting. Faster emergency response and disaster aid (saving lives, reducing injuries). Enhanced healthcare delivery via telehealth and real-time health monitoring (especially for elderly/disabled). More accessible city services for all (IoT-enabled assistance for disabled, multi-language support, etc.). Increased civic engagement and transparency (open data, citizen feedback loops)	Ex: Singapore's nearly crime-free year after IoT camera rollout (public safety); A disabled person using a smart traffic signal app to safely cross (inclusion); Kansas City's open data portal letting residents track city performance (engagement)

(continued)

Table 2.3 (continued)

Sustainability Pillar	G-IoT Contributions and Benefits	Examples from Smart Cities
Economic (efficiency, growth, innovation)	- Cost savings in city operations (energy, water, maintenance) free up budget for social programs New business opportunities and jobs in the green tech and IoT sector (economic growth with sustainability focus). Attracting residents and businesses with improved city services and quality of life (competitive advantage). Data-driven optimization leads to productivity gains (e.g., less time in traffic, efficient logistics)	Ex: Smart building energy savings translating to millions of dollars saved and reinvested in affordable housing; Startups emerging to use city open data to build new services (economic innovation); A city noted for its smart infrastructure drawing tech companies and talent, boosting the economy (while adhering to sustainability values)

It's clear that G-IoT's impact on social sustainability is multi-faceted. Ultimately, a smart community is one where people feel safer, more connected, and more empowered to shape their environment. IoT provides the digital connective tissue to achieve this, but human-centered design and governance are key—the technology must be deployed in ways that address community needs and protect rights (privacy, security, equity). When done right, the city of the future is not just one that is high-tech, but one that is greener, healthier, and more liveable for everyone.

2.7 Enhancing Urban Sustainability with G-IoT to Promote a Circular Economy

Rethinking Urban Sustainability and Circular Economy Principles

Sustainability in cities isn't only about daily operations; it's also about rethinking how resources flow through the urban economy. The concept of the circular economy has gained traction as a way to minimize waste and keep products and materials in use for as long as possible (as opposed to the traditional linear "take-make-dispose" model). G-IoT plays a pivotal role in enabling circular economy strategies in cities by providing tracking, automation, and data that help close the loops of resource use.

2.7.1 Smart Waste Management and Recycling

Cities generate vast amounts of solid waste daily. Traditional waste management often ends in landfills or incineration, which are not sustainable. IoT is revolutionizing this sector in several ways, making it more circular (by increasing recycling and recovery) and more efficient:

- **Intelligent Collection Systems**: Waste bins with IoT sensors facilitate on-demand collection based on fill levels, minimizing unnecessary truck trips and emissions. Various waste streams, such as recyclables and organics, can be collected separately as needed, increasing recycling rates. Recycling rates in some cities have increased by 30–50% and contain less contamination due to timely pickups. Moreover, AI-based route optimization reduces operational expenses by as much as 50% and reduces carbon emissions, such as Sydney's 20% CO_2 decrease. Reduced truck trips also make the air cleaner and cities healthier.
- **IoT-Enabled Recycling Bins and Sorting**: Urban areas are implementing smart bins in public spaces that employ sensors and AI to automatically recognize and sort trash into the proper bins, reducing contamination in recyclables. While still in development, this technology has great promise. In recycling plants, more sophisticated AI systems with high-speed sensors and robotic arms sort materials by type, shape, or brand—frequently employing QR codes or digital stamps. These industrial IoT solutions significantly enhance sorting accuracy and efficiency over manual processes, enabling more materials to be recycled. IoT data also informs recycling markets and policy decisions.
- **Waste-to-Energy and Resource Recovery Optimization**: For the waste that isn't recycled, many cities use waste-to-energy (WTE) plants or biogas digesters (for organic waste). IoT sensors in these facilities can optimize the process—for example, monitoring the calorific value of incoming waste to adjust combustion, or tracking the composition of biogas to tweak digestion parameters. Additionally, IoT can monitor emissions in real time to ensure these facilities operate cleanly. While WTE is sometimes controversial, if used, IoT makes it cleaner and more efficient, extracting maximum energy while minimizing pollutants. In a circular mind-set, even ash or digestant from these processes can be reused (e.g., ash in construction, digestant as compost) if carefully monitored for safety.

2.7.2 Organic Waste Composting (Closing the Loop on Food and Green Waste)

Organic waste (food scraps, yard trimmings) is a major part of city waste and a significant source of methane (a potent greenhouse gas) if landfilled. The circular approach is to compost or anaerobically digest this organic material and return it to soil or use it as biofuel. IoT helps manage composting at both industrial and community scales:

- **Industrial Compost Facilities**

Industrial composting facilities employ IoT to track important conditions such as temperature, moisture, and oxygen levels in compost piles. Real-time information is transmitted through wireless sensors to central systems, enabling operators to correct dryness or overheating issues by adding moisture or aerating piles. This facilitates quicker, more efficient composting, minimizes odors and emissions, and enhances quality of the compost. Some systems even aerate automatically through blowers, assisting in the transformation of waste into useful compost for application in parks or agriculture—closing the nutrient cycle.

- **Community Composting and Smart Bins**

Locally, IoT is being implemented in household and neighborhood composting systems. Compost bins that are "smart" are able to monitor fill levels and conditions of compost and alert users when intervention is necessary. A company from Australia created a tool that provides instant feedback through an application, facilitating easier and more interesting composting. A few cities provide kitchen bins that use IoT and monitor the amount of compost, relying on feedback and gamification to make people more interested. These initiatives divert organic waste away from landfills, enrich soil, and sustain urban agriculture, encouraging circular resource use.

2.7.3 Digital Product Passports and Lifecycle Tracking

The most revolutionary impact of IoT to drive a circular economy is its ability to create digital product passports (DPPs)—information records accompanying a product that indicate its source, materials, maintenance history, and end-of-life treatment. These are enabled through technologies such as RFID tags, NFC chips, and QR codes.

- **Unique Identification and Tracking**

Each product or part can be labeled with an IoT-readable unique ID, and it can then be tracked throughout production until it is disposed of or reused. For example, a washing machine with an internal RFID tag will be able to hold information regarding its parts and repair history. When it has reached the end of its lifecycle, recyclers can scan and determine what material it contains as well as the disassembly protocols, assisting recycling properly. Customers in secondary markets can also read its service record, encouraging good reuse.

- **Internet of Things and Blockchain for Supply Chain Transparency**

Merging IoT with blockchain delivers secure, tamper-proof product tracking along every step—factory, store, consumer, and recycler. Each IoT scan records to a blockchain, creating a reliable record. The EU is already exploring this for products such as electronics and cars. Such systems not only optimize recycling effectiveness but also stimulate reuse—for instance, scanning an expired item to learn its repair details and reuse possibility. DPPs retain material in use by educating all involved parties.

- **Lifecycle Performance Data**

IoT can track usage and wear over time, enabling circular business models. For instance, sensors on public scooters can identify repeated component failure, prompting manufacturers to enhance durability. In such models as lighting-as-a-service, where the owner is the service provider, IoT enables monitoring of usage, maintenance initiation, and return for reuse or recycling at the end of service life—all promoting circular design and asset management.

2.8 Smart Waste Management and Urban Mining

IoT and data analytics are opening new paths to the recycling of valuable resources from city trash—transforming cities into "urban mines." By adding technology such as X-ray or laser sensors to sorting facilities, cities can track trends in waste, like increased e-waste in certain districts. This information can inform specialized recycling initiatives or laws. Sensors may also identify valuable metals, such as copper or gold, bound for landfills, which could make it worth investing in advanced material recovery. IoT-enabled tags on electronics can also simplify tracking and separation when recycling.

- **Facilitating Circular Systems**

One of the biggest challenges of circular economies is coordinating the behavior of producers, consumers, and recyclers. IoT fills those gaps by giving them real-time information. It assists manufacturers in creating improved, longer-lasting products and provides consumers with more intelligent disposal choices via apps and smart bins. Those devices can also encourage green behaviors—if a person can see that they recycled 10 kg of plastic this month, they're more likely to continue doing so, just as fitness apps encourage repeated use.

- **Beyond Waste: Closing the Loop**

Circularity is not just limited to solid waste—it involves water and energy as well. IoT systems can handle greywater reuse or harness waste heat from industries to heat buildings, reducing resource loss. The aim is to establish feedback loops that emulate natural ecosystems, where waste turns into input.

References

1. Cook, D., & Davíðsdóttir, B. (2021). An appraisal of interlinkages between macro-economic indicators of economic well-being and the sustainable development goals. *Ecological Economics, 184*, 106996.
2. Alsamhi, S. H., Ma, O., Ansari, M. S., & Meng, Q. (2019). Greening internet of things for greener and smarter cities: a survey and future prospects. *Telecommunication Systems, 72*, 609–632.
3. Bibri, S. E. (2023). Data-driven smart eco-cities of the future: an empirically informed integrated model for strategic sustainable urban development. *World Futures, 79*(7–8), 703–746.
4. Martikka, M., Salo, S., Siilin, K., Ruohomäki, T., Tuomaala, P., & Nykänen, E. (2018). Smart city resilience with active citizen engagement in Helsinki. In *2018 International Conference on Intelligent Systems (IS)* (pp. 162–167). IEEE.
5. Ariyachandra, M. M. F., & Wedawatta, G. (2023). Digital twin smart cities for disaster risk management: a review of evolving concepts. *Sustainability, 15*(15), 11910.
6. Tan, D., Setiawan, J., Anela, S., & Rustian, S. A. (2024). Lessons Learned from Building Smart Cities: A Systematic Literature Review of Case Studies in Singapore and Seoul. *G-Tech: Jurnal Teknologi Terapan, 8*(4), 2505–2513.
7. Knieps, G. (2017). Internet of Things and the economics of smart sustainable cities. *Competition and Regulation in Network Industries, 18*(1–2), 115–131.
8. Garg, P., Pranav, S., & Prerna, A. (2021). Green Internet of Things (G-IoT): A Solution for Sustainable Technological Development. In *Green Internet of Things for Smart Cities* (pp. 23–46). CRC Press.
9. Chaudhari, B. S., Zennaro, M., & Borkar, S. (2020). LPWAN technologies: Emerging application characteristics, requirements, and design considerations. *Future Internet, 12*(3), 46.
10. Kenway, S. J., Lant, P. A., Priestley, A., & Daniels, P. (2011). The connection between water and energy in cities: a review. *Water Science and Technology, 63*(9), 1983–1990.
11. Wilson, L., Lichinga, K. N., Kilindu, A. B., & Masse, A. A. (2021). Water utilities' improvement: The need for water and energy management techniques and skills. *Water Cycle*, 2, 32–37.
12. Hashmi, M. H. S. M. K., Hänninen, S., & Mäki, K. (2011, October). Survey of smart grid concepts, architectures, and technological demonstrations worldwide. In *2011 IEEE PES conference on innovative smart grid technologies Latin America (ISGT LA)* (pp. 1–7). IEEE.
13. Goyal, M. K., Kumar, S., & Gupta, A. (2024). AI Innovation for Water Policy and Sustainability. Springer Nature Switzerland
14. Frana, P. L. (2024). Assessing Smart Nation Singapore as an International Model for AI Responsibility. *International Journal on Responsibility, 7*(1), 2.
15. Ivanova, M. (2020). Coloring the UN environmental: The catalytic role of the UN environment programme. *Global Governance: A Review of Multilateralism and International Organizations, 26*(2), 307–324.

AI and the Green Internet of Things (G-IoT) in Smart Cities

3

The rapid rise of urbanization in the twenty-first century requires urgent actions from the world's cities to come up with effective and sustainable ways to manage resources, infrastructure, and services. Smart city is the product of modern technology such as AI, G-IoT to maximize the urban life function. AI and G-IoT are collaborating to provide the foundation for developing sustainable and smart urban systems due to their capability of handling large data volumes and efficiently operating energy-saving IoT systems. And with AI technologies, technological applications can give cities more ability to make their decisions fact based in urban development and public transportation systems. G-IoT Tangible urban infrastructure also benefits from G-IoT whose devices are smart enough for applications to thrive with the least environmental clutter. Cities that integrate these technological systems create smart connected environments, where such operations can be sustainable according to the visions of sustainability and technological development. To elaborate more on the impact of AI on smart cities, [1] evaluates current infrastructure development and urban utilization and transport, and sustainability and local authorities' operations and so much more. The conversation shows how AI solutions are put into practice across cities across the globe and explores the impact of technology on urban life. An important current aspect of the chapter considers integration challenges, such as data privacy risks, cyber threats, and system interoperability and it provides some guidance in terms of how to mitigate these challenges. AI and G-IoT are the core mode of urban construction to improve operation efficiency, reduce energy consumption, and improve living quality of residence. The chapter concludes with a deep-dive analysis of these achievements that will serve city managers and technology policymakers and lawmakers in constructing a framework of a smarter sustainable city.

V. P. Gupta et al., *Green IoT and AI for Sustainable Development of Smart Cities*,
Synthesis Lectures on Sustainable Development,
https://doi.org/10.1007/978-3-031-98743-4_3

3.1 Introduction

Urban technology progressions via Artificial Intelligence and Geo-IOT promote fast growing of smart cities foster the life quality on neighborhoods at cities. Artificial Intelligence can also help a large city's operations become more sustainable and efficient while providing better living conditions. AI in conjunction with G-IoT contributes in better asset management decisions and environmental conservation (e.g., both support energy-efficient IoT devices). This section explores convergence of AI technology and G-IoT systems for building future urban environment enhancement, while focussing on four smart city aspects; traffic management systems and public safety measures, environmental observation and, energy-efficient infrastructure development of urban landscapes. Combining G-IoT leads to power conservation devices that emit less carbon from urbanization. AI and G-IoT together build up a sustainable system that adapts to the need of current urban society which sees the combination of the advanced technologies and the concern toward the environment. Cities are provided with copies through digital twinning, where time sensitive viewing is assisted for the observation, the prediction of subsequent events and testings in case of emergency. AI-based environment monitoring systems promote sustainable solutions by monitoring pollution trends, tracing greenhouse gas emissions and making recommendations for remediation to better protect the environment. AI adoption also brings benefits to public safety and security operations. With predictive analysis, the patterns of criminals can be predicted beforehand to provide prevention in the first place. Automated diagnostic systems in combination with robotic surgery on top of devices that can monitor our body in real time will make medical care more accessible for everyone!

3.1.1 The Role of AI in Shaping Smart Cities

The twenty-first century has turned cities into dynamic epicenters of economic development during their expansion characterized by innovation and cultural diversity. Fast urban development brings a series of problems and challenges, such as: traffic congestion, resource consumption, environment deterioration and imperfect public service facilities. Smart cities will only succeed in achieving their capability of a sustainable and efficient city by the utilization of advanced technologies such as G-IoT and AI. Sustainable power consumption for devices in inspire the AI's capability of fast analysis of big data hence smart strategies makes cities smarter and greener by G-IoT. This aspect focuses on the transformation of urban environment and public services delivery, together with improvement of the quality of life of the inhabitants as a result of AI based interventions. In this section, we show how G-IoT is interconnected to AI in order to give rise to cities, urban development planning and infrastructure control for public services delivery and transportation improvements, enhanced energy utilization etc. The following section discusses the key challenges cities face during the implementation of AI-based technologies, and presents solutions to surmount such obstacles, such as in data protection, user privacy, and technology compatibility issues.

3.1.2 The Evolution of Smart Cities: The Role of AI

Smart cities are the urban ecosystems of smart cities, and the modern urban systems runs on data, leading to the need for increased efficiency and sustainability levels using AI. Classic cities acquired information through humans decisions and managed everything by hand, creating delays and inefficiencies on decision making and resources placement. Artificial intelligence consists of machine learning algorithms, deep learning and predictive analytics in order to automate decisions completely and optimize the operation of city management.

(1) AI as the processing unit for smart cities for smart cities, AI systems serve as the processing unit, collecting real-time data from sensors as well as cameras and IoT devices. These cognitive computation systems pattern spot while you make future sales estimates his or her own and more and automations that can cause cities operate more nearly as good as they must. AI-powered traffic control systems automatically change traffic signals to prevent gridlock by predicting where there will be congestion in the future and making those predictions accordingly.
(2) Digital Twins Assisted by AI is the most advanced application of AI-enabled urban planning. By using the data from the AI algorithms, for buildings and transportation (as well as infrastructure components), virtual copies of them can be generated, which facilitate being able to perform the simulations in multiple situations. Urban Planning—Digital twins The application of digital twins in urban planning Digital twins in Urban Planning allows urban planners to identify sustainable practices, perform impact assessment of the projects and to improve the municipal processes.
(3) Internet-based decisions reinforce the basis of urban governance from the analytical perspective. Predictive analysis mashup is used for the cities to be able to mitigate disaster and avoid the crisis as well as in allocating resources efficiently. The AI analytical system crunched crime data in order to create locations for where police need to focus their attention activities.

3.1.3 The Synergy Between AI and G-IoT

Using the G-IoT sustainability model, we minimise the power related measurements needed by Iot devices protecting the environment. The combination of AI and G-IoT is able to offer the following efficiency gains to smart city management:

G-IoT equipped devices G-IoT equipped devices having smart meters in combination with energy-aware sensors as well as intelligent lighting systems employ AI algorithms for optimal utilization of power. Maximum power consumption rates are predicted to rearrange the distribution of power, with less-waste products and less-environmental emissions.

In contrast, G-IoT products use energy-efficient electricity from a data system and return the emission control recommendations to those regulators.

With Smart trash Management AI, G-IoT together it becomes less time consuming to collect your trash and trucks are optimally routed for efficiently punctual services where the amount of fuel needed to power these trucks is minimized. AI-based trash segregation systems make meaningful enhancements in the recycling rate for improving the environment condition of urban life for us to be sustainable.

3.1.4 AI Applications in Smart Cities

There are many sectors in smart cities where the industry relies on AI, such as transportation, energy management, public safety and healthcare. AI effects on three critical faculties are as follows:

Smart Transportation Systems

Real-time alterations in traffic light routines by AI-based data analyses aid to reduce traffic jam in the roads. Schools teach students about self-driving cars with AI taking the wheel to free up public transportation and eliminate driver-driven mistakes and increase drive safety. Using predictive analytics AI delivered public transport systems generate improved bus and metro timetable schedules, therefore reducing waiting times.

AI-Enhanced Public Safety

AI surveillance systems analyse video data to identify abnormal behavior patterns, which are then sent to the authorities live in police departments. The combination of predictive policing and AI emergency systems, for disaster, provides more effective disaster monitoring systems.

Sustainable Infrastructure

AI is being used to make building management more efficient, with HVAC control systems, lighting automation and environmentally friendly building construction. AI predictive analysis designs, as used in programmes, will reduce our operations costs and extend the operational life of our city infrastructure systems.

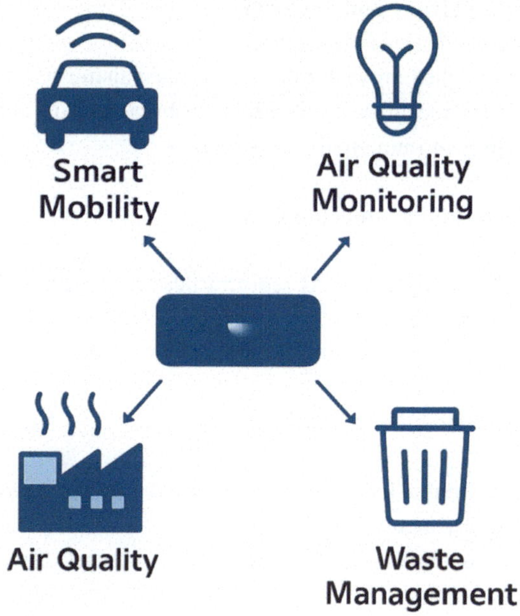

Fig. 3.1 Application of AI and big data in smart cities

Artificial Intelligence in Governance and Citizen Services
FAEs which are AI-driven virtual assistants assist citizens by engaging them to quickly fetch information from the Government and also provide access to easy issue reporting. Individuals, with these AI-run e-governance systems, get more transparency and more of a voice in their communities.

AI in Healthcare
While smart cities are becoming part of the landscape of both developed and developing countries, the IoT technology for enabling smart cities has been extended to the health-care area, i.e., in the form of smart health-care. Smart telemedicine with AI-enabled technology and predictive diagnostic tools and real-time patient diagnostic systems along with patient tracking systems work in smart cities to provide better access to medical services. The integration of AI analytics makes for better predicting of pandemics and preparing for response solutions (Fig. 3.1).

3.1.5 Challenges in Implementing AI in Smart Cities

AI in smart cities, however, despite its potential to be transformative, faces several challenges:

Data Privacy and Security

- AI relies on vast database stores, raising fears about privacy of personal data.
- Cybersecurity issues like data breaches and manipulation of A.I. represent vulnerabilities in municipal infrastructure.

High Implementation Costs

- The adoption of AI requires heavy investment in infrastructure, sensors and processing power.
- Tight budgets limit the expansion of AI programs in poor cities.

Interoperability of Technology

- Various AI systems and IoT devices need to work seamlessly across different municipal departments.
- Standardized procedures are necessary for data exchange between AI-connected systems.

Ethical Considerations

- AI decision making should be transparent and without prejudice.
- Regulation should ensure that AI applications are ethical and not discriminatory.

3.1.6 Future Prospects of AI in Smart Cities

As the AI technology gets better, the future smart cities will also get more independent, self-sufficient and resilient. Some rising trends include:

AI-Driven Climate Resilience

- A.I. algorithms predict natural disasters and optimize climate adaption strategies.

Hyperconnected Smart Cities

- AI will support real-time interoperability between a whole range of urban systems, so that smart, city-wide automation becomes easy.

AI on Quantum for Future City Control

- Quantum computing will enhance the capacity of AI, facilitating faster and more complex decision-making.

AI-Integrated Smart Grids

- Power networks that reflexively adjust the flow of energy based on demand at any given moment.

Conclusion

AI and Green Internet of Things (G-IoT) are transforming smart cities, emphasizing efficiency, sustainability, and better human quality of life. AI-driven decision making, smart infrastructure, smart mobility and sustainable energy solutions are redesigning the cities of the future. While challenges such as data privacy, ethical concerns, and technical integration are obstacles to overcome, continued innovation and policy-based AI governance will usher in a new era of smart, green, and connected urban landscapes. AI combined with G-IoT will significantly contribute to the future smarter and responsive cities which are environmentally sustainable for future generations [2].

3.2 AI-Driven Infrastructure and Urban Planning for Revolutionizing City Development

3.2.1 Introduction

Running with this idea of city planning meant relying to a much greater extent on manual data collection and more traditional modeling techniques. AI disrupts and redefines city development with its ability of real-time data processing and predictability, deployment, and automation and data rich analysis and analytics. For example, AI-based Geographic Information Systems (GIS) can offer a dynamic mapping functionality allowing planners to simulate urban sprawl and perform environmental analysis. G-IoT temperature developing air quality and noise sensors and water quality Establishes this set of capabilities shows the staff to continued Monitoring as a manager of the city. AI-driven smart grids add more efficiency to power grid networks so that cities can grow in a sustainable manner.

It can be safe said that AI-driven urban planning should be now classified as generative design as a major breakthrough according to Hadiyana and Ji-hoon [3]. An AI algorithm system is used to explore massive numbers of design scenarios to find the sweetest spots for everything from roads, as well as sustainable placements of buildings and green space. By integration of AI based intelligent energy distribution model and

Fig. 3.2 AI-driven infrastructure

AI based waste management system optimal recycling analysis for reducing the electricity outage is also provided. AI-powered infrastructure is heavily dependent on Building automation systems (BAS) as one of its critical enablers. BAS with AI and IoT sensors control home and office heating and cooling systems so that they use as little energy as possible. Applying AI-based construction monitoring AI is an effective tool to enforce safety rules by identifying and possibly defects as early as possible and minimum any detrimental delays and costly down-time (Fig. 3.2).

With the spread of Artificial Intelligence (AI) technologies, cities are planned, developed, managed, and maintained by modern automated ways, and in a world fully filled with huge data, AI-enabled systems are synchronized with advanced IoT networks, real-time analytics, and intelligent infrastructure to create efficient, resilient, and sustainable urban environments. AI driven solutions help optimize city planning and infrastructure use, anticipate trends in the growth of cities, and improve city resilience, adaptability, and sustainability to emerging challenges. In these conditions AI is awarded an unparalleled head start in multidisciplinary life-data-integration, predictive analytics, machine learning models, and automation procedures for the infrastructure design and the urban planning. AI in urban planning helps local governments accomplish better decision-making efficiency, decision speed, and decision quality, thereby optimizing land use, constructing energy-efficient buildings, and planting smart transit systems, and resilience cities.

3.2.2 The Role of AI in Urban Planning and Smart Infrastructure

AI and its promise of data-driven decision-making "Contemporary planning practice has a lot to gain from the use of artificial intelligence, particularly when it comes to urban development, as it enables cities to use data to make informed decisions. Here are the primary methods in which artificial intelligence supports the application of intelligent infrastructure and the advancement of urban planning:

1. Predictive Analytics of City Growth.
2. Geospatial analysis powered by AI predicts urban development trends, guiding planners to invest resources judiciously.
3. Satellite imagery coupled with AI can review past trends and extrapolate variations in population density, letting communities predict where more homes, transit and utilities will be needed.
4. Land-use efficiency is evaluated using machine-learning algorithms that ensure urban expansion causes the least amount of environmental impacts, while ensuring sustainable growth.
5. AI-Driven, Smart Infrastructure Services and Solutions 76.
6. AI enhances the efficiency of city infrastructure by predicting maintenance of wear and tear in roads, bridges, and buildings.
7. Computer vision and sensor-based AI models analyze real-time structural health data, so that cities can circumvent infrastructure problems before they rise up.
8. AI-driven traffic management systems optimize road use by adjusting traffic signals and providing alternative routes to minimize congestion.
9. AI in City Planning for Sustainable Solutions
10. AI enables the construction of energy-saving buildings, such as intelligent HVAC systems, smart lighting and automatic energy management technology to minimize power usage.
11. With AI-based green space design, urban sustainability is reinforced such as ensuring the right place for parks, water reservoirs or tree plantations contributing to minimize urban heat consequences.
12. AI-powered waste management systems improve trash pickup, recycling and waste management, efficient and helping cities become more sustainable.

3.2.3 The Impact of AI on Transportation and Mobility Infrastructure

Transportation and mobility systems play crucial role for efficient transport of people and goods in urban areas. Here's how AI-powered innovations are changing the shape of transportation infrastructure:

AI-Based Optimization of Traffic Flow

Traffic prediction models relying on artificial intelligence process historic as well as present data and work on better utilization of road resources to reduce congestion.

Intelligent traffic lights managed by AI change on their own based on the flow of traffic, eliminating congestion and reducing fuel consumption.

AI-enabled public transportation scheduling ensures buses, trains and metros run at maximum efficiency, reducing the waiting time and providing a better customer experience.

Autonomous, Intelligent Transportation Systems

- AI is ultimately clearing the way for autonomous vehicles of all types, self-driving buses, taxis and delivery trucks.
- AI-based ride-sharing and on-demand mobility solutions contribute to the efficiency of urban mobility via a decrease in the rate of individual automobiles and an optimization of vehicle usage.
- Smart parking solutions that use AI, also known as AI powered smart parking systems, guide cars to open parking spots, saving fuel and preventing traffic buildup.

Sustainable Transportation Networks Empowered by AI

- AI via bicycle-sharing and e-scooter sharingzi hope to reduce reliance on fossil fuels, while promoting green commuting.
- AI models assess pedestrian movement to create safer and more walkable cities at the intersection of smart crosswalks and pedestrian-prioritized areas.
- Maximize fuel economy and reduce your carbon footprint with AI-based fleet management in mass transit.

3.2.4 Smart Infrastructure for Resilient and Disaster-Ready Cities

And catastrophe preparation and resilience is one of the key parts of AI in city planning. The contribution of AI-enabled technologies to city resilience is shaped by:

AI FOR DISASTER PREDICTION AND MANAGEMENT

- A.I.-enabled early warning systems crunch seismic, meteorological and environmental data to predict natural disasters like earthquakes, floods and hurricanes.
- Emergency response coordination through AI ensures that rescue is efficient and well-organized, resulting in fewer deaths and damages.

AI for Climate-Responsive Urban Planning

- "AI-enhanced climate models prepare cities for negative weather by modeling out future situations and instructing us to build infrastructure."
- Two AI-based smart water system applications are employed to continuously identify drought and to optimize water distribution for sustainable urban water supply.
- AI-empowered renewable energy grid management improves solar and wind energy usage, and helps make cities more climate-resilient.

3.2.5 Challenges and Future Prospects of AI-Driven Urban Development

Challenges:

Data Privacy and Security: The accumulation and processing of large amounts of urban data raise concerns about data security and citizen privacy.

High Capital Expenditure: AI based smart infrastructure requires huge money which limits its penetration in developing world.

Integration Challenges: Cities will need to ensure that AI-powered products and services integrate seamlessly with our current networks and governance.

Ethical issues: The decision-making by AI should be transparent, just and fair, so as to reduce urban bias and injustice.

3.2.5.1 Future Prospects

1. AI Integrated Smart networks: Cities will transition to fully AI-linked power networks that enable real-time optimisation of power distribution.
2. Quantum AI for Urban Development: Quantum computers will enable far more sophisticated simulations and predictive modeling for urban planning.
3. (AI-Driven) Hyper-Connected Cities: Ubiquitous AI connecting all elements of the city will enable self-operating and self-sufficient city formations.

Conclusion

AI is revolutionizing urban infrastructure and city planning and the way forward is more sustainable, resilient and efficient smart cities. With AI-powered predictive analytics, real-time automation, and resilient design, cities can transform use of land, improve public transit, minimize natural resources impact, and quality of urban life. Yet, new challenges about high prices, or about data privacy have to be tackled, but AI-driven urban planning seems to have a bright future, providing cities that are smarter, greener and more adaptable to urban dwellers' ever-changing needs.

3.3 Challenges and Key Requirements for Smart Cities

3.3.1 Introduction

Although AI and G-IoT show great potential in smart city construction, there are also some problems to be solved. These are the data privacy concerns, the infrastructure costs, the technical compatibility and the security risks. AI systems require vast amounts of data to work well, raising ethical and privacy concerns around data collection and use.

In addition, incorporation of AI in the current infrastructure of a city, implies huge financial investment. Cities have to set rules to ensure fair access to smart technology and to maintain cybersecurity so that people's information is secure. Furthermore, interoperability between different AI systems and with IoT devices is a basic requirement for a hassle-free realization of smart cities. Data fragmentation is one of the key challenges in AI-enabled smart cities. Cities are getting data from many places, from traffic cameras to environmental sensors and utility meters, but there's no platform for integrating all that data together. AI Data Lakes and decentralised ledger technologies like blockchain, offer interesting possibilities for improved data integration, transparency and security. A second challenge is the ethical application of AI. In AI systems, bias can result in biased decision making in applications such as facial recognition and predictive policing. To mitigate this, AI governance models, such as explainable AI (XAI) methodologies, should be enforced to ensure a degree of transparency and accountability for smart city applications [4].

There are faces number of barriers and key demands that we must overcome in order to ensure that we make that step to an AI-powered sustainable urban future. Although much potential exists for AI and G-IoT to make contributions to infrastructure performance improvements, public services provision, and the development of smart cityscapes, there are various barriers to their realisation. The challenges are including data privacy issues, high infrastructure expenses, compatible technologies, and risks of cyber security which are barriers for more extensive application. In addition, if cities are to be able to integrate the AI benefits effectively, they need to meet a certain set of criteria—such as solid governance frameworks, scalable technological infrastructure, and a focus on sustainability.

This section will detail the most significant barriers and unmet needs in the move towards efficient, sustainable and intelligent smart cities including stakeholder analysis of ethical AI application, data security, affordability, policy frameworks and public participation.

3.3.2 Key Challenges in Smart City Development

Privacy and security aspects of data

Big data from AI-driven smart city applications will give rise to deadly privacy and security problems.

The AI-connected surveillance, smart grid and intelligent traffic systems all require data constantly collected raising the specter of unwanted access and data breaches.

Solutions such as blockchain, end-to-end encryption and strict data governance regulations are crucial to ensure secure AI usage.

Higher Costs of Implementation and Maintenance

Smart city projects require substantial funds to be poured into AI-supported infrastructure, cloud services, IoT gadgets and cyber security approaches.

Most poor cities have issues with finance and cannot bring in high-end AI solutions at scale.

Public and private partnership cooperation is required to address the funding gap and obtain sustainable funding for smart city projects.

I. Technical Challenges

AI-powered smart cities are based on several sets of technology, such as the IOT sensors, cloud platforms and AI analytics.

Absence of common AI integration frameworks results in incompatibility issues between different vendors and city departments.

The use of open-source AI models and common communication protocols is key to ensure seamless interoperability.

Ethics and Bias in AI Decision Making

Both AI systems used in urban planning and design and for law enforcement and public services need to be fair, objective, and ethical.

Decision making that is biased and discriminatory is also a potential consequence of unethically trained algorithms on AI.

Transparent AI governance principles and ethical AI development strategies are needed to ensure the fairness and equity of smart city services.

Public resistance and ignorance.

- Lots of people are wary about A.I. surveillance systems, data-collection methods and robots replacing human jobs.
- A lack of education of the benefits of AI in smart cities is contributing to public opposition to new technology.
- Cities need to promote engagement, transparency and learning activities to foster trust in AI-based urban development.

I) Cyber-threats and AI frailties

- AI-powered smart city systems are vulnerable to cyberattacks, hacking and data manipulation.
- AI-powered predictive maintenance and traffic control networks can be breached causing disruption in critical services.
- Governments need to develop holistic cybersecurity standards, AI-based threat detection mechanisms and AI governance mechanisms to mitigate risks.

3.3.3 Key Requirements for Smart City Success

Scalable and adaptive AI infrastructure

- Smart cities will need to increasingly invest in scalable, cloud-based AI tech to accommodate growing urban populations and tech advancements.
- The predictive analytics that are powered by AI should be adaptable enough to meet the changing needs of the city ensuring sustainability over time.

Regulatory Environment and Policy.

- Governments should establish concrete rules about AI ethics, data security and responsible AI deployment.
- Policies must foster transparent data collection, AI decision making free from bias and ethical AI governance.

Inclusion of Environmentally Friendly Technologies

6.1.2 Sustainability becomes the fundamental concept in an AI-enabled smart city, including renewable energy power, green internet of things (IoT) devices, and low-carbon AI models.

AI-enabled energy management systems allow intelligent electricity usage with less common urban carbon foot prints.

PPPs on a cooperative basis with the public sector

Effective smart city investment requires cooperation between governments, technology firms and educational bodies.

Public–private partnerships (PPPs) can ease fiscal limitations and stimulate the advent of AI enhanced urban planning.

Smart City Development: A Citizen Point of View

- Smart cities with an AI focus should focus on improving citizen well-being, access and inclusion.

Table 3.1 Challenges and key requirements for smart cities

Challenge	Key Requirements
Traffic congestion	- Intelligent traffic management systems - Real-time public transport data
Energy inefficiency	- Smart grids - Renewable energy integration - Energy usage analytics
Waste management	- IoT-enabled waste bins - Route optimization for collection - Recycling tech
Water scarcity	- Smart water metering - Leak detection systems - Water reuse infrastructure
Air pollution	- Air quality monitoring sensors - Emission control policies
Data security and privacy	- Strong data governance - Encryption and secure communication protocols
High implementation costs	- Public–private partnerships - Government subsidies - Scalable solutions
Digital divide	- Universal internet access - Digital literacy programs
Urban population growth	- Scalable infrastructure - Efficient urban space utilization

- Success will require participatory governance techniques, aI-boosted public feedback systems and transparent municipal management (Table 3.1).

Although AI and G-IoT can turn urban environments into highly effective, sustainable, and intelligent smart cities; their realization involves significant challenges. Problems like data protection, cost constraints, interoperability, ethical considerations, and cybersecurity threats need to be addressed for a successful implementation of smart city. But with scalable AI infrastructure, good governance mechanisms, incorporation of green technology and public–private partnerships in place, cities can harness the full potential of AI-driven urban transformation. The vision for smart cities hinges on a balanced approach that fosters technology-driven innovation and guarantees sustainability, inclusiveness and safety for all citizens.

3.4 Smart Transportation Systems and Mobility Solutions Using AI

Smart cities are built on smart transportation and AI will change urban transportation. Transportation offerings driven by AI help to better manage traffic flows, reduce congestion and increase the efficiency of public transport. We use state-of-the-art machine-learning algorithms to predict traffic, which we can use to modify your route in real-time and avoid waiting. AVs are another example of the significance of AI in the modern transportation. Pilotless cars that make use of AI sensors and deep learning algorithms to navigate large metropolitan areas safely. G-IoT contributes to the overall sustainability of transportation by enabling the operation of energy efficient vehicles, reducing vehicle emissions, and facilitating the convergence of charging infrastructure and smart grids for electric vehicles.

AI also contributed to the development of smart public transport systems. Smart buses and trains employ AI to evaluate passenger data, improve timetables, and deliver real-time updates, boosting commuter experience. Predictive maintenance enabled by AI protects the lifetime of transport infrastructure, decreasing downtime and boosting service dependability.

Beyond conventional transport systems, AI is changing urban air mobility (UAM). AI-powered drone deliveries, flying taxis, and hyperloop systems are being explored to offer quicker and more efficient travel alternatives. AI-based mobility-as-a-service (MaaS) solutions unify diverse transport modes—ride-sharing, public transportation, and bike rentals—into a seamless customer experience (Fig. 3.3).

Transportation is one of the most crucial components of a smart city, directly impacting economic activity, sustainability, and the quality of life for citizens. The integration of Artificial Intelligence (AI) in transportation systems is transforming urban mobility, optimizing traffic flow, decreasing congestion, boosting public transit efficiency, and enhancing road safety [5]. Real time data, predictive analytics, machine learning algorithms and automation are combined in AI-powered smart transportation systems to create frictionless, smart mobility networks.

In this part of the course, we will see how AI is changing urban transportation, focusing on self-driving cars, smart traffic management, AI-driven public transit, and sustainable mobility. Furthermore, it will discuss the challenges and opportunities related to the use of AI in the transport system, such as legal issues, security, privacy, and readiness of the infrastructure.

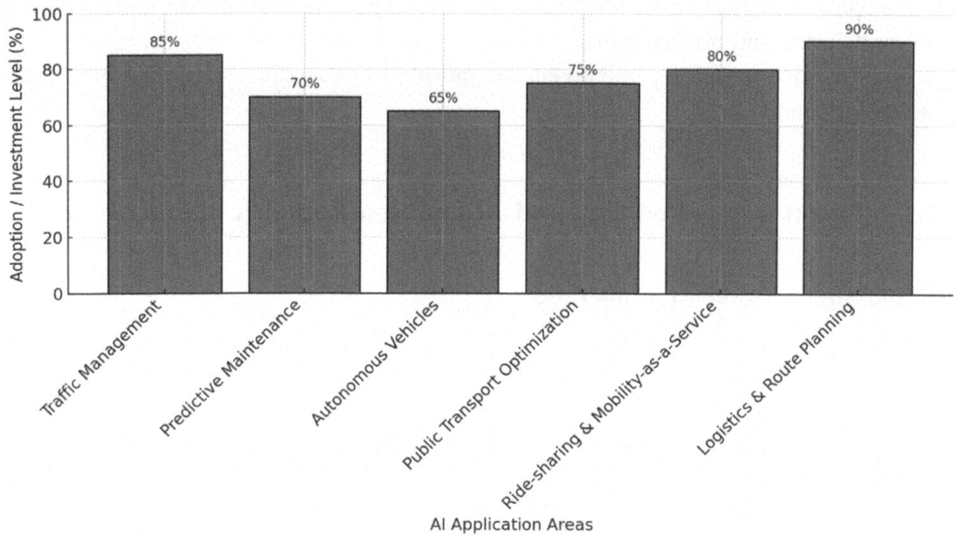

Fig. 3.3 AI applications in smart transportation systems and mobility solutions

3.4.1 AI-Driven Traffic Management Systems

Smart Traffic Light Control

1. AI-enabled traffic light systems use real-time sensor data and predictive analytics to improve traffic light timing and ease traffic.
2. Algorithms that use machine learning algorithms analyze traffic patterns and adjust lights on the fly to optimize flow and reduce wait times.
3. Such systems contribute to reduction in fuel consumption and emissions, and thus promote an ecological urban transportation.

PFT: Predictive Flow Traffic

1. AI uses past traffic patterns and live GPS data to pinpoint areas of congestion and suggest alternative paths.
2. Predictive models allow municipal planners to be more proactive with regard to congestion control solutions.
3. Smartphone apps backed by AI provide commuters with up-to-the-minute traffic times and route recommendations.
4. Incident Detection and Response Based on AI.
5. Computer vision and AI-powered analytics detect traffic accidents, road hazards and erratic driving behaviors on-road.

6. Automated emergency control systems can be warned without human intervention to re-route traffic and prevent jams.
7. It is imperative to develop intelligent monitoring to drive road safety and emergency acting response.

3.4.2 Autonomous Vehicles and AI-Enabled Public Transportation

Cars with Brains and Self Driving Cars

- Self-driving vehicles with the power of AI employs deep learning, LiDAR and computer vision to safely navigate through urban streets.
- Smart cities are building dedicated holes in the ground, in the roadway and elsewhere, to aid self-driving cars.
- Computer-directed driving saves fuel, reduces human error and improves road safety generally.

AI for Public Transit Optimization

- AI-powered bus and metro systems use real-time passenger data to enhance timetables and improve efficiency.
- AI-powered smart ticket systems enable contactless payments and dynamic pricing models.
- Chatbots and virtual assistants driven by artificial intelligence enhance customer experience in public transportation systems.

3.4.3 Sustainable and Green Mobility Solutions

AI in Smart and Shared Mobility

- AI has been used to drive electric vehicle (EV) adoption by optimizing charging station locations and the relationship between EVs and the grid.
- App-based ride-sharing services propelled by artificial intelligence slash car idling time and carbon footprints.
- Smart Hubs in mobility: At the intersection of AI, IoT and renewable energy to create organized, sustainable transportation systems.

Intelligent Infrastructure for Bicycles and for Pedestrians

- Artificial intelligence urban planning tools optimize bike-sharing schemes and pedestrian paths.
- AI-powered sensors enhance street lighting and pedestrian safety, and support sustainable, non-motorized mobility.

3.4.4 Challenges and Future Outlook

While AI transport ecosystems come with numerous benefits, they also face significant barriers, including concerns around data privacy, cybersecurity, regulatory and legal compliance, and the capability of national infrastructure to adapt. Private tech companies and academics need to work with governments to create safe, ethical, and scalable AI-driven transportation solutions. As AI progresses, so will cities, birthing a new age of connected, efficient, sustainable urban transportation for generations to come.

3.5 Leveraging AI for Sustainable Urban Living with Energy-Efficient Buildings

Sustainability is one of the cornerstones of smart cities, and AI is instrumental in the development of energy-efficient urban living spaces. AI-enabled smart buildings integrate automated systems for controlling lights, heating, and cooling when and where people occupy them, thus minimising the need for energy.

IoT-connected green sensors track building energy consumption and identify inefficiencies while suggesting remedial measures. So-called smart grids, fuelled by artificial intelligence (AI), will be developed to better balance supply and demand of electricity, which will continue to increase the incorporation of renewable energy by the grid, reducing waste. The same developments result in less carbon footprints and savings in money for both individual people and companies.

Moreover, AI gives ability of predictive analytics and sustainable resources control. Having analysed weather conditions and past records, AI models can maximize the use of water and power at urban homes. Intelligent waste management solutions help to increase recycling, reduce landfill rubbish, and keep a circular economy within communities.

The increasing urban population and resource consumption has driven the development of eco-cities focusing on energy savings and eco-environment through technological solutions. Buildings (about 40% of global energy consumption) are of great importance with respect to urban sustainability. Combining AI and energy-efficient buildings is a game-changer and it empowers creative solutions for reducing energy, carbon emissions, and increasing the building performance overall.

In this section, we will focus on AI technologies are changing the way energy efficient buildings of smart cities used to work. It will discuss topics like intelligent HVAC systems, AI for energy management, predictive maintenance, real-time monitoring and the impact of IoT in building automation. The discussion will also address challenges in implementing AI-based solutions as well as the future of sustainable urban living.

Table 3.2 outlines various AI technologies applied in the transportation and mobility sector. It demonstrates AI-based use cases, including predictive traffic analytics, autonomous vehicles, and smart parking solutions. For each application, the table describes the role of AI (such as machine learning or computer vision), the benefits (like alleviating congestion or increasing efficiency) and include real-world examples of companies or platforms that employ these technologies, including Google Maps, Tesla and Uber. The broad theme here is the role of AI in improving efficiency, safety, convenience, and cost-effectiveness in transportation.

3.5.1 AI-Driven Energy Optimization in Buildings

AI-Powered Smart HVAC Systems

1. HVAC systems are powered by AI and use real-time data to adjust temperature, airflow, and humidity on the fly.
2. AI predicts the perfect heating and cooling cycles so as to not waste any energy, and make for a more comfortable environment.
3. AI-powered smart thermostats respond to users and external temperature factors to minimize energy consumption.
4. Systems and Prediction Based Energy Management.
5. Real-time energy monitoring (AI) finds inefficiencies, and provides corrective measures.
6. Predictive analytics predict future energy requirements, so you can implement energy-saving solutions before demand gets ahead of consumption.
7. AI-based automation helps tailor lighting, ventilation and electricity consumption to the occupancy requirements and the external environment.
8. AI for Integrating Renewable Energy.
9. With AI, you get the most from solar panels, wind turbines, and battery storage for max efficiency.
10. AI-supported smart grids control energy distribution, using less non-renewable sources.
11. Artificial intelligence based demand response system control building's energy utilisation in response to grid availability and price in real time.

Table 3.2 Smart transportation systems and mobility solutions using AI

Technology	Application	AI Role	Benefits	Real-World Examples
Predictive traffic analytics	Traffic forecasting, congestion management	AI models analyze historical + real-time data	Reduced congestion, optimized traffic lights	INRIX, Google Maps
Autonomous vehicles (AVs)	Self-driving cars, delivery robots	Computer vision, ML algorithms for navigation and control	Safety, efficiency, cost reduction	Waymo, Tesla
Smart parking systems	Real-time parking availability	Sensors + AI detect occupancy, predict availability	Reduced time to park, reduced emissions	ParkWhiz, Bosch smart parking
AI-powered public transport	Bus/train route optimization, dynamic scheduling	AI algorithms optimize schedules and passenger loads	Increased efficiency, improved passenger experience	Singapore LTA, Moovit
Mobility-as-a-service (MaaS)	Integration of multi-modal transport on one platform	AI recommends optimal routes and pricing	Convenience, cost-effective travel	Uber, Citymapper, Whim
Demand forecasting	Ride-hailing, public transport planning	AI forecasts peak times and rider demand	Better resource allocation, shorter wait times	Uber, Ola, Lyft
Intelligent traffic signals	Adaptive signal control systems	AI adjusts signals in real-time based on traffic flow	Reduced travel time, fuel savings	Siemens Sitraffic, Surtrac (Pittsburgh)
AI in logistics and fleet Mgmt	Delivery routing, vehicle health monitoring	AI optimizes routes, predicts maintenance needs	Faster delivery, lower operational costs	FedEx, Amazon logistics

3.5.2 Challenges and Future Prospects

So, the benefits of the AI empowered energy efficient building are evident, but issues like high deployment cost, data privacy, and integration challenges are still inhibiting them of being mainstream. However, the development of AI, IoT and green technology has been increasingly opening the door for such solutions to be more available, more reasonably priced, and more cost-effective in the end.

AI, and renewable energy and smart infrastructure are key to the future of sustainable urban living, where these factors will work in harmony. As cities implement AI-based energy management, their overall efficiency and environmental footprint as well as the quality of life for citizens will improve. With AI shaping society's urban future, and today's pressing urban issues demanding sustainable solutions, I predict AI will be the leading influence on the development of our future smart cities.

3.6 Optimizing City Operations for Better Quality of Life with AI-Enabled Services

AI-powered innovations go beyond infrastructure and sustainability and significantly enhance the quality of life in smart cities. AI-powered health services also enhance patient surveillance, predictive diagnostics and emergency responses. AI-connected wearable IoT devices can also make it possible for real-time monitoring of health, thereby making it easier to sense any health problem in the initial stage and treat it soon.

Governance based artificial intelligence drive more the public administration by streamlining provision of services like garbage management and disposal, public security surveillance and community participation model. AI chatbots and virtual assistants are available for people to easily navigate government services, including mitigating the country's bureaucratic bottlenecks (Fig. 3.4).

Rapid urbanization worldwide has resulted in a growing demand for sustainable, effective, and pleasant urban environments. Cities today are wrestling with complex problems in transportation, health care, energy use, garbage disposal and public safety as they grow. To address these challenges, smart technology specially AI and Green Internet of Things (G-IoT) have been rapidly adopted in cities. Not only improving the operation and management of the city but also enhancing the overall living conditions of citizens. This chapter examines how the integration of AI and G-IoT can enhance urban operations, and sustainability, and grow the intelligence and livability of cities.

G-IoT and AI Integration in Smart Cities

In the smart city setting, AI and Green-Internet of Things (G-IoT) are considered as key technologies in changing the urban life when they are being integrated together. The G-IoT means the group of interconnected things that want operate in an eco-energy manner. With the integration of sustainability features to include low energy usage and

Fig. 3.4 Optimizing city operations for better quality

efficient resource management G-IoT facilities are changing the way cities manage and monitor their urban infrastructure.

AI, however, enables data-driven decision making, automation and optimization by being able to learn from huge amounts of this data coming in from these interconnected devices. When AI is integrated into the G-IoT system, it is able to analyze and process this data to improve how urban facilities function on the spot. From this connectivity comes the potential to develop smarter, more sustainable cities, which can respond to dynamism dynamic challenges—think changes in energy demand, congestion in transportation patterns and inefficiencies in waste handling.

For instance, AI-based applications can analyse IoT-enabled sensor data in smart buildings to optimise heating, ventilation, and air conditioning (HVAC) systems, reducing energy consumption while maintaining comfort. In the same vein, AI technology can process information from sensors monitoring traffic to optimize traffic light periods leading to a reduction in congestion and atmospheric carbon release. In these interfaces, G-IoT and AI complement one another to offer better urban lives promoting energy, sustainability and ease.

Integration of AI and G-IoT to Control and Manage Smart City Using Machine Learning Algorithm

This hybrid AI-G-IoT-ML has the possibility to reveal new and fresh solutions for coping with the complexity of the modern cities. Artificial intelligence, a subfield of AI, comprises the creation of algorithms that can learn from and make predictions or decisions based on data without being explicitly programmed. Considering the smart city, ML algorithms can analyze input from G-IoT devices regarding trend identification, prediction-making, and operations refinements.

Predictive Maintenance Predictive maintenance is one of key applications of ML algorithms in smart cities. AI-enabled systems can detect anomalies and predict potential breakdowns before they take place by processing data that has been gathered by IoT sensors placed within infrastructure like public transportation systems, bridges and highways. This type of preventative maintenance reduces downtime, lowers repair bills and ensures that vital infrastructure remains safe and reliable.

Another application is in energy systems. When AI is combined with G-IoT, it enables the citywide real-time monitoring of energy use and makes it possible to balance supply and demand. For instance, IoT enabled smart grids can monitor energy consumption to track energy usage and AI algorithms can predict periods of peak demand and automatically adjust power distribution to mitigate shortages or waste. These solutions help not only in optimizing energy usage, but also in reducing the carbon footprint of cities, contributing to the broader sustainability goals of smart cities.

In public transport, the convergence of AI and G-IoT provides for better optimal routing and scheduling. Real-time IoT device data including GPS trackers and passenger counts combined with machine learning algorithms could predict traffic patterns, change routes on the fly and provide passengers with real-time travel information. This increases the reliability and efficiency of public transportation, and reduces congestion and pollution (thus improving the user experience).

Energy Consumption Perspective: Green IoT (GIoT) and Its Role

A leading benefit for the Green Internet of Things (G-IoT), is its ability to help IoT devices to save energy. Conventional IoT devices commonly require large amounts of energy in order to operate, especially when it takes place at an urban scale and in a city wide location. This can result in higher energy use, operating cost and environmental load. Alternatively, G-IoT is intended to address these challenges by integrating power saving techniques and reducing the environmental footprint of IoT devices in general.

G-IoT devices commonly adopt energy-efficient wireless access technologies, e.g., Low Power Wide Area Networks (LPWAN), and in-the-field energy-supplying devices (EH), to reduce their energy consumption. They can operate for long periods without charging or replacing batteries, so they are ideal for long-term installation in smart cities. For example, in smart waste management systems, sensors could gauge fill levels of garbage bins and transmit the information to central systems without wasting tremendous amount of electricity. And smart streetlights with motion sensors could adjust their brightness based on up-to-the-minute activity, dimming when the streets are empty and increasing energy usage only when someone is on the block.

Further, with G-IoT, cities can control and optimize energy use in various sectors like at home, in industry etc. With IoT sensors to scrutinize the use of energy, cities can find that there are areas of waste and can take specific actions to reduce it. Such things as smart buildings, in which IoT-based energy management systems monitor lighting, heating, and cooling based on patterns of occupation, bring about less energy use without loss of comfort.

Learning-by-Doing via Smart City Development Case Studies from Real Life

This book allows learners to experience real-world case studies in the development of a Smart City. Case studies can provide the opportunities for understanding how the application of the AI + G-IoT may contribute to improving the municipal operation and the quality of life of city ecosystems. From this overview of smart city approaches throughout the world, readers can learn valuable lessons on overcoming challenges, and best practices to use and pitfalls to avoid when deploying smart solutions.

For example, the city of Barcelona has emerged as a leading paradigm for a smart city that has deployed so many different IoT and AIdriven solutions to smarten up municipal government. Smart lighting, waste management, smart water meters—all of these innovations have contributed to a massive reduction in energy consumption, better public services and higher sustainability in Barcelona. An equally large city, Singapore, has relied on AI and G-IoT to optimize traffic, enhance public safety, and facilitate urban planning.

These cases also provide empirical lessons on how AI and G-IoT could be used to address challenges specific to different cities. They emphasize that in building smart cities it is essential that there would be collaboration between the government, private sector companies, and citizens [6]. They also underscore the importance of evidence-based decision-making, real-time monitoring and forward-looking management in building sustainable and liveable urban environments.

Cross-disciplinary Concepts in the Design of Smart Cities

Smart cities build process is interdisciplinary bringing to together various fields like urban planning, engineering, information technology and environmental science [7]. Collaboration among a variety of stakeholders, including city planners, data analysts, policymakers and technology developers is needed in the application of AI and G-IoT for urban management. This interdisciplinary composition allows us to develop full solutions that can meet the complex requirements of new urban context.

For instance, city planners might partner with AI experts to create smart transport networks that alleviate traffic congestion and reduce carbon emission. Engineers may work with G-IoT developers to build power-conservative buildings and infrastructures. Environmental scientists could help ensure that smart city programs align with sustainability goals by evaluating the environmental footprint of new technology.

Realisation of smart cities would require vision with collaboration cutting across disciplines. By fostering collaboration and embracing diverse ideas, cities can maximize the opportunities offered by AI and G-IoT to create urban landscapes that are efficient, sustainable, and adaptable.

The Role of Technology and Sustainability in Smart City Initiatives

Fusion of technology and sustainability is a big component of smart cities development. As cities expand, the urban landscape is facing an increasing onus to balance growth with environmental protection. AI and G-IoT significantly contribute in maintaining this balance, which can enable cities to optimize resource utilization, minimize waste, and promote environmental sustainability.

For instance, AI-based energy management systems could reduce the carbon footprint of cities by improving energy usage in buildings, transportation, and public facilities. G-IoT systems can help cities surveil air quality, water consumption and waste management, providing essential information that leads to informed decisions about environmental protection.

It's essential to bring sustainability to the concepts and practices of smart city, including sustainable information systems that can keep cities liveable for the next generations. Prioritizing sustainability, and by leveraging the potential of AI and G-IoT, cities are able to build smarter, greener, and more resilient urban landscapes.

Conclusion

AI and the Green Internet of Things (G-IoT) AI and G-IoT are changing the way cities operate, offering innovative strategies to address urban challenges, promote sustainability and enhance quality of life for the citizens. By integrating these technologies, cities can minimize energy use, raise public services levels and create more sustainable and attractive urban environments. Featuring real-world case studies and cross-disciplinary ideas, the book is a useful resource for readers with an interest in the practical applications of AI and G-IoT in smart city construction and is a guide for all readers wishing to use these technologies to build better cities.

3.7 AI Applications in Smart Cities in the Real World

There are many examples of AI powered smart city initiatives being implemented successfully by cities across the globe. In Singapore, Barcelona, and New York City, artificial intelligence is being employed to manage traffic, optimize public housing, and even deliver on the promise of 'predictive' policing to improve urban life. Disasters: AI and disaster response AI disaster response capabilities also highlight the transformative role of AI in creating resilient and adaptive cities.

The idea of smart city has emerged as a disruptor of urban transformation leveraging on the possibilities offered by new technologies to improve urban life. Artificial Intelligence (AI) is a key part of this transformation, with solutions to enhance a sustainable, efficient and an enhanced quality of life. AI also enables smart cities to be more responsive to the ever-changing needs of its residents through detailed data analysis, process automation, as well as real-time decision-making. This chapter explores the applications of AI in smart cities, analyzing how AI-based technologies are transforming urban management, enabling more sustainable, efficient, and livable environments.

3.7.1 Energy Management and Optimization

Energy management is among the largest applications of AI in smart cities. Big cities devour enormous amounts of energy, in buildings or transportation, lighting, and countless other demands that all come with feet. AI can further enhance energy distribution and utilization by processing live data from smart meters, sensors, or IoT.

For instance, within financial services AI systems make projections about energy usage that help utilities better anticipate supply and demand and within smart grids, forecasts energy-consumption, enabling utilities to align the supply of energy with demand. During times of high demand, AI could decide to allocate energy to critical infrastructure and reduce energy use in less important locations. AI can also control energy storage, allowing renewables such as solar and wind to be used while cutting reliance on fossil fuels. AI is also involved in demand side management by offering users visibility of how they consume energy and proposing the best ways to reduce it.

The likes of Amsterdam have already successfully implemented AI within its energy system to make its smart grids more efficient and reduce overall carbon footprint. By optimizing energy usage with AI, cities can accelerate the transition towards a more sustainable and resilient energy future.

3.7.2 Traffic and Transportation Management

The most widespread problems of cities are the traffic jam and the inefficientness of public transportation. AI is transforming traffic and transportation management in smart cities when it connects with IoT devices and sensor networks.

For example, traffic control systems that run on AI could analyze information from sensors and camera and GPS technology installed in vehicles to adjust the timing of traffic signals, ensuring the smooth flow of traffic and preventing backups. This prevents traffic jams, reduces fuel consumption and lowers air pollutants." In cities such as Singapore, AI has been employed to make predictive traffic systems which have helped increase the flow of traffic in rush hour, slashed journey times and limited vehicle emissions.

Furthermore, AI enhances the rise of driverless cars that is purposed to revolutionize urban transport. Self-driving vehicles which rely on AI can connect to smart city networks, increasing safety on the roads and reducing the human intervention required for traffic management. This technology, when implemented, will also help in reducing the accidents, improve the efficiency and build a greener transport ecosystem.

Management and recycling of waste

AI techniques in smart cities are also contributing to transforming the management of rubbish. Conventional methods of collecting garbage have low efficiency and lead to wasted materials and power. But when combined with IoT sensors, these AI-based solutions offer improved waste management systems that not only optimize collection schedules but also monitor trash levels and ensure the efficient use of resources.

AI-equipped sensors fitted in trash bins can measure fill levels and relay real-time data to waste management staff in many locations. This information enables teams to refine the timing and routes for collection, with less fuel wasted on inefficient journeys. AI is also powering smart waste disposal intiatives in some cities like in Barcelona where the technology is used to monitor the amount of garbage, enabling more efficient pickup and preventing garbage overflow in public spaces.

AI is also helping cities increase recycling rates by automating sorting functions. Algorithms AI can analyze what's in the trash and identify items that can be recycled and sorted accordingly. This has not only the advantage of optimizing the recycling process, but also contributes to sustainabilty by reducing the number of garbage being shipped off to the waste dumps.

3.7.3 Urban Planning and Infrastructure Development

AI is enabling smarter and more responsive urban planning through access to data-driven insights to make better informed decisions [8]. Algorithms can sift through huge pools of information from various sources (satellite pictures, traffic records, census data, and so on) to help planners make informed decisions about infrastructure development, land use and public services.

For instance, AI might help to see where people are overpopulating and predict where development is going to occur in the future and to tell us where to be building more infrastructure, like roads, schools, and hospitals. AI can also help design and construct buildings and public spaces so that urban spaces are functional, environmentally friendly, and aesthetically pleasing.

In places like New York City, AI can help urban planning projects determine the impact that a new development will have on the environment or how riders will react to a new bike lanes or where to place new buildings. Being able to predict future urban trends and enhance urban plans is essential for building cities capable of accommodating growing populations while minimizing their environmental impact.

3.7.4 Public Safety and Security

AI tech is helping to enhance public safety and security in smart cities. Armed with machine learning and computer vision capabilities, it can analyze data from surveillance cameras, sensors and other monitoring systems to detect and predict potential security threats on a continuous basis.

AI-guided surveillance: In cities like London, surveillance technology with AI can identify abnormal behavior or suspect activity which processes the camera information and raises an alert. Such technologies could potentially warn the police of potential threats, before anything occurs. AI can also be employed to predict crime trends by analysing historical crime data, so that police forces can allocate resources more effectively and take preventative action in high-risk areas. AI is also enhancing reduced emergency response time. With the integration of AI and live traffic insight, pubic safety agencies can ensure that their ambulances, fire trucks, and police cruisers are dispatched and adequately maneuvered to incidents more efficiently and more quickly than before. AI-powered drones are also used to assist in search and rescue operations—providing real-time information from disaster sites and allowing first responders to make faster decisions.

3.7.5 Water and Environmental Monitoring

Artificial intelligence (AI)-enabled technologies are being applied more and more for the monitoring and to manage rubbish, water quality and for resolving environmental issues in smart cities. Monitoring Atmospheric, Soil, and Water Quality AI technology can process data from sensors in water, air, and weather stations to provide real-time views on the state of the environment. This empowers cities to quickly respond to environmental challenges—such as pollution, droughts and floods.

In Tokyo, AI systems analyze data from sensors in rivers, lakes and reservoirs to track water quality and detect pollution, for example. These AI tools can predict potential contamination events, enabling water treatment facilities to take preemptive measures and ensure that communities have access to safe water. Likewise, air monitoring systems using artificial intelligence (AI) technology can alert occupants to unhealthy levels of pollution and allow them to take informed actions on their health and personal activities.

AI is also working to help manage natural resources by enhancing irrigation systems, tracking soil quality, and aiding in sustainable farming efforts. AI through data driven insights enables cities to more effectively monitor their environmental resources, and meet long run sustainability goals.

3.7.6 Citizen Engagement and Services

AI is changing the dialogue between cities and their inhabitants, and making these services more convenient, responsive, customized. Chatbots, virtual assistants, and self-service platforms powered by AI are allowing constituents to interact with their local authorities and access services with greater ease and efficiency.

For example, AI-based platforms could allow citizens to report issues such as potholes or broken streetlights or water leaks—making sure the immediate service requests are routed to the right department and addressed quickly. AI can also help communities customize their services according to their residents' needs with tailored recommendations on public transit, trash collection and healthcare services.

In Helsinki and other cities, AI is even used to provide citizens with up-to-the-minute information about public services: updates on shifts in public transit schedules or data about emergency services. Underpinning communication and interaction, AI ensures that cities are increasingly responsive to the needs of the people who live in them, improving satisfaction and quality of life.

Conclusion

AI is transforming how cities are governed, offering more efficient, sustainable and pleasant environments. From energy saving and traffic control to smart garbage collection, public safety and citizen engagement, AI is helping the creation of smarter cities that can help solve the problems of the future. Cities are beginning to realize the opportunity AI can provide to improve quality of life, streamline services and ensure resources are utilized effectively through practical implementation. The future is bright as AI has potential to boost the productivity and sustainability of cities, and there are many possibilities for urban expansion in the future.

Smart Cities' fusion of AI and G-IoT in smart cities implies that by the next 10–12 years, a new form of structuration of urban development has been proposed. AI-based infrastructure, intelligent transportation, energy-saving buildings, and efficient municipal management help realize sustainable urban life. Although challenges such as privacy of data and interoperability of technologies exist, relentless advancement in AI and G-IoT will shape smart cities where we live, work, and play and where everything is sustainable, efficient, and green.

References

1. Sustainable Living (2024). Food and Industry 5.0: Transforming the food system for a sustainable future. Springer.
2. Rani, S., & Taneja, A. (2024). WSN and IoT: An integrated approach for smart applications. Springer.

3. Hadiyana, T., & Ji-hoon, S. (2024). AI-driven urban planning: Enhancing efficiency and sustainability. Springer.
4. Farhaoui, Y., Bhushan, B., Sindhwani, N., Anand, R., & Luh, A. (2024). Artificial intelligence and machine learning in smart city applications. Springer.
5. Khang A. (2025). Driving Green Transportation System Through Artificial Intelligence and Automation. Lecture Notes in Intelligent Transportation and Infrastructure. https://doi.org/10.1007/978-3-031-72617-0.
6. Sowmya, D. S. (2024). Case studies and real-world applications. In Smart city applications: Trends and challenges (pp. 345–360). IGI Global.
7. Riffat, S., Ahmad, M. I., & Shakir, A. (2024). Eco-cities: Sustainable urban development. In Green technologies for sustainable urban planning (pp. 123–145). Springer.
8. Mitra, S., Madhuvanthi, S., & Sabumon, P. C. (2024). Nature-based urban planning strategies. In Innovations in sustainable urban development (pp. 89–110). Springer.

5G Networks Towards Sustainable Smart Cities

4

4.1 Introduction to the Next Generation of Connectivity: 5G Technology

5G technology, the fifth generation of cellular wireless technology, aims to revolutionize connectivity by offering significantly faster speeds, lower latency, and greater network capacity compared to previous generations like 4G [1]. This translates to transformative potential for various industries and applications, including autonomous vehicles, smart cities, and immersive experiences.

The introduction of 5G technology marks the beginning of a revolutionary age that will drastically alter the connectivity landscape and open up previously unheard-of opportunities for the construction of sustainable smart cities. This paradigm shift establishes the foundation for an intelligently regulated and completely integrated urban ecology, surpassing the incremental advancements of preceding generations by a significant margin. The crucial infrastructure underpinning for the realization of resource-efficient and ecologically mindful urban settings is 5G [2].

5G is a complete technological revolution that includes three main pillars: Massive Machine-Type Communications (mMTC), Ultra-Reliable Low-Latency Communications (URLLC), and improved Mobile Broadband (eMBB). It is not just about better internet speeds. Together, these developments enable a wide range of applications essential to sustainable smart cities.

- **Enhanced Mobile Broadband (eMBB)**: eMBB allows for the smooth transfer of massive amounts of data by offering noticeably faster data rates. Applications like high-definition video surveillance, augmented and virtual reality (AR/VR) for civic involvement and urban planning, and the effective provision of digital services all

V. P. Gupta et al., *Green IoT and AI for Sustainable Development of Smart Cities*,
Synthesis Lectures on Sustainable Development,
https://doi.org/10.1007/978-3-031-98743-4_4

depend on this capability. 5G is intended to "support high data rate applications and provide a better user experience" through its eMBB capabilities [3].

- **Ultra-Reliable Low-Latency Communications (URLLC)**: By reducing latency, URLLC makes communication almost instantaneous. For time-sensitive applications like industrial automation, remote surgery, and driverless cars, this is essential. As [4] highlighted, URLLC is a "disruptive technology direction for 5G" that enables "mission-critical applications". For systems that need real-time responsiveness to operate safely and effectively, it is imperative to be able to guarantee minimal data transmission delays.
- **Massive Machine-Type Communications (mMTC)**: The implementation of sensor networks for smart waste management, smart grids, and environmental monitoring is made possible by mMTC's support for connecting a large number of IoT devices. As [5] emphasized, the "Internet of Things for smart cities" relies on the ability to connect a "massive number of devices". This capability is fundamental for gathering the granular data necessary for intelligent urban management.

Synergistic smart city development is driven by a collaborative ecosystem formed by the integration of 5G, Green IoT, and AI. 5G's high bandwidth and low latency enable real-time data acquisition and transmission from IoT sensors. AI algorithms process this data to automate processes, predict trends, and optimize resource allocation. Such synergy facilitates easier creation of responsive, intelligent, and environmentally sustainable urban environments.

4.1.1 Detailed Comparison of 4G and 5G

Feature	Explanation	5G Performance	4G Performance	Real-World Applications
Higher data speeds	Data transmission speeds are greatly increased by the upcoming 5G network, allowing for faster downloads and continuous streaming	Can achieve speeds up to **20 Gbps**	Generally supports speeds up to **1 Gbps**	Instead of taking many minutes, a full-length 4 K movie may be downloaded in a matter of seconds Without any interruption, with clear streaming in ultra-HD video without any buffering

(continued)

(continued)

Feature	Explanation	5G Performance	4G Performance	Real-World Applications
Reduced latency	5G reduces data transmission and reception latency, greatly increasing the effectiveness of real-time communication	Can reduce latency to 1–2 ms	Typically experiences 20–70 ms delay	• Real-time reaction in competitive online gaming • Smooth video conferencing with no discernible lag • Quicker trade execution in capital markets where milliseconds count
Greater device capacity	5G networks are perfect for expanding the Internet of Things since they can accommodate a far greater number of connections at once	Can manage up to 1 million devices per square kilometer	Handles around 100,000 devices per square kilometer	• Sensor-based smart city infrastructure that controls and reduce the uses of energy use and also control traffic • The extensive use of IoT in industrial automation • Smart houses with AI-powered appliances that are fully connected
Stronger network reliability	For mission-critical applications, 5G guarantees a more reliable and continuous connection	Offers highly reliable connectivity	Can be less reliable in high-traffic areas	• Autonomous cars that need continuous, real-time connectivity for safety and navigation • Robotic procedures that are controlled remotely and require no lag • Uninterruptible, safe military communication networks

Source Author's creation

4.1.2 New Applications Enabled by 5G

Application	How 5G enables it	Real-world examples
Autonomous vehicles	5G's ultra-low latency and high-speed connectivity allow real-time communication between vehicles, infrastructure, and pedestrians, ensuring safe and efficient self-driving operations	• Self-driving cars that react instantly to traffic signals and road conditions • Smart traffic management reducing congestion and accidents • Vehicle-to-vehicle (V2V) and vehicle-to-infrastructure (V2I) communication
Smart cities	5G supports widespread deployment of sensors and connected devices that collect and analyze real-time data to improve urban infrastructure, public safety, and energy efficiency	• Intelligent traffic lights adjusting to real-time traffic flow • Smart waste management systems optimizing garbage collection routes • AI-powered security surveillance and emergency response systems
Immersive experiences (VR/AR)	The high bandwidth and low latency of 5G allow seamless virtual and augmented reality experiences, enabling real-time rendering of high-quality graphics	• AR navigation apps that superimpose mobile screens in real time • Virtual travel, which allows shoppers to tour locations remotely • VR-based training for engineers, surgeons, and pilots
Remote surgery	5G ensures instant communication between doctors and robotic surgical equipment, allowing highly precise remote operations with real-time haptic feedback	• Surgeons performing operations from miles away using robotic arms • Telemedicine advancements improving healthcare access in remote areas • AI-assisted surgeries with real-time data analysis
Industrial automation	5G enables smart factories by connecting machinery, sensors, and AI systems, allowing real-time monitoring, predictive maintenance, and automated production	• AI-powered robotics assembling products in real time • Predictive maintenance reducing machine downtime • Real-time quality control using AI-driven inspection systems
Mobile cloud gaming	5G's ultra-fast speeds and low latency enable smooth, high-resolution gaming without the need for high-end hardware, as processing happens in the cloud	• Streaming games on mobile devices without lag or downloads • Multiplayer gaming with real-time responsiveness • AR-based mobile games with instant interaction and updates

Source Author's creation

4.1.3 Specific Applications and Their Impact

- **Sustainable Transportation**: 5G makes it possible to have intelligent traffic management systems that maximize traffic flow, minimize congestion, and lower emissions. Connected and autonomous cars, with 5G's low latency backing them up, can transform urban mobility, making people less dependent on private cars and encouraging public transport usage.
- **Environmental Sustainability**: Real-time information regarding air and water quality, sound levels, etc., is obtained through 5G-enabled sensor networks, and this facilitates environment management in real time. Efficient energy consumption is ensured through optimized smart grids assisted by the low latency and high bandwidth of 5G and encouraged integration with renewable energy.
- **Effective Resource Management**: 5G accommodates smart water management systems that track water consumption and identify leaks, minimizing wastage of water. Smart waste management systems streamline waste collection and recycling, minimizing landfill waste.
- **Improved Public Safety**: Surveillance systems and emergency response networks made possible by 5G improve public safety and security. Real-time data analytics and AI-based applications facilitate quicker and better responses to emergencies.
- **Smart Agriculture**: In cities, through vertical farming, and roof gardens, 5G, IOT and AI can be employed to regulate water use, and pesticide use. This will assist in growing local food, and lowering the carbon footprint involved in long distance food transport. [6] emphasizes the significance of "Edge AI for 5G heterogeneous ultra-dense networks" which is ideal for these city farming projects.

Challenges and Considerations

While 5G is full of promise, there are challenges to its deployment. 5G infrastructure deployment require both meticulous planning and a significant financial investment. Due to the massive amounts of data generated by 5G networks and the increasing number of connected devices, cybersecurity is a significant concern. Creating inclusive and equitable smart cities involves ensuring digital equity. In addition, the energy consumption of 5G networks must be managed with caution to ensure the technology promotes sustainability goals instead of undermining them. Among the key challenges is "balancing performance and energy efficiency in 5G wireless networks," [7]. Shafi et al. [8] also identify the need to pay serious consideration to the "requirements and challenges" of "5G for future vertical industries."

In short, the future generation of sustainable smart cities is based on the pillar of 5G technology. Its enhanced capabilities, including eMBB, URLLC, and mMTC, enable a range of applications that increase productivity, reduce their adverse impacts on the environment, and enhance the quality of life of the residents. Cities can create more resilient, sustainable, and livable cities by overcoming the deployment challenges and leveraging its transformative power.

4.2 Key Features and Capabilities of 5G Technology

A key milestone in wireless communications development is the introduction of 5G technology. 5G's architectural innovations and features extend beyond merely delivering higher data rates; they are designed to accommodate the next generation of networked systems, particularly in the context of sustainable smart cities. In an effort to assist in the development of effective and environmentally aware urban spaces, this aims to explore the major aspects of 5G required to facilitate the application of green IoT and AI technology [6] (Fig. 4.1).

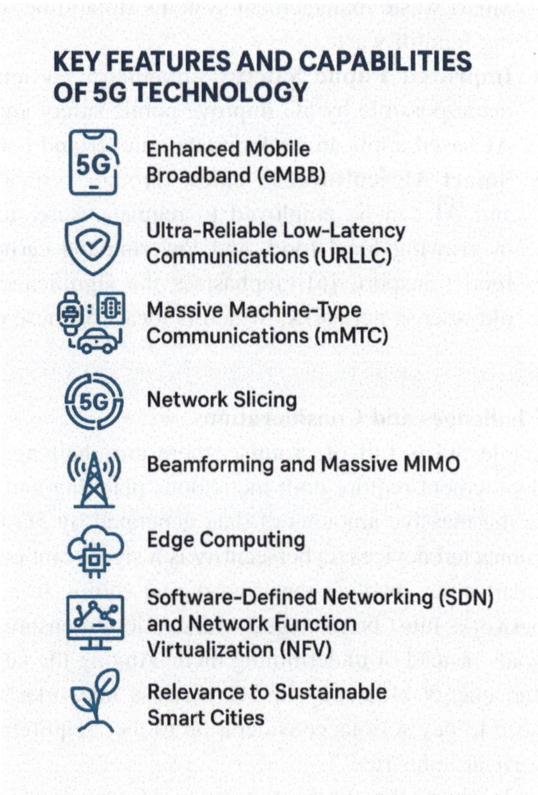

Fig. 4.1 *Source* Author's creation

4.2.1 Enhanced Mobile Broadband (eMBB): Driving High-Bandwidth Applications

5G's eMBB feature is one of its strongest aspects. This provides significantly higher data rates compared to previous generations, supporting bandwidth-intensive applications essential to the operations of smart cities in the modern era. Such applications as immersive augmented and virtual reality experiences for public services, remote health-care consultations in real time, and high-definition video monitoring of traffic for traffic control purposes are facilitated. Based on studies, 5G can deliver user-experienced data rates of 100 Mbps and peak data rates of 20 Gbps, which means that these data-intensive applications can run smoothly [9].

4.2.2 Ultra-Reliable Low-Latency Communications (URLLC): Ensuring Critical Operations

A fourth key element of 5G is Ultra-Reliable Low-Latency Communications (URLLC), meant to enable applications that need ultralow latency and high reliability. Applications such as remote surgery, industrial automation, and autonomous vehicles take huge advantage of URLLC's 99.999% reliability and 1 ms end-to-end latency [10, 11].

4.2.3 Massive Machine-Type Communications (mMTC): Connecting the IoT Ecosystem

Most IoT devices can be made connected owing to 5G's Massive Machine-Type Communications (mMTC) support. This feature renders big-sized sensor networks ideal for asset tracking, metering, and environmental monitoring. mMTC supports one million devices per square kilometer, which renders it crucial to build end-to-end smart city infrastructure based on data-centric decision-making [12].

4.2.4 Network Slicing: Tailoring Networks for Diverse Needs

Network slicing allows for the deployment of virtualized and isolated network partitions that are designed for each service and application separately. This feature can be exploited by smart cities to give priority to essential services such as emergency response and public safety by assigning certain network resources. 3GPP standards facilitate network slicing by offering compatibility and flexibility [10, 11].

4.2.5 Beamforming and Massive MIMO: Enhancing Signal Efficiency

Beamforming and Massive Multiple-Input Multiple-Output (MIMO) technologies enhance signal quality and coverage by pointing radio waves at targeted users or devices.

This is very beneficial for green IoT deployment because it enhances network efficiency, reduces interference, and has lower power consumption.

As per [13], massive MIMO can significantly enhance bandwidth usage and spectral efficiency.

4.2.6 Edge Computing: Enabling Real-Time Processing

The other essential element of 5G is edge computing, which shifts data processing and storage towards the network edge. Edge computing enables real-time applications and computationally intensive activities by minimizing latency and bandwidth limits. Split-second decisions for video analysis and autonomous vehicles, for instance, are required. Edge computing reduces latency since it processes data locally, closer to the point of data.

4.2.7 Energy Efficiency: Supporting Sustainable Deployments

In order to lower energy use, cutting-edge features like adaptive power control and sleep modes have been built into the 5G architecture. This matters to smart, green cities in that it integrates green concepts into ICT infrastructure. Checko et al. [14] point out that 5G networks have been planned to consume less power than 4G networks, particularly in low-traffic environments.

4.2.8 Software-Defined Networking (SDN) and Network Function Virtualization (NFV): Facilitating Network Flexibility

Software-Defined Networking (SDN) and Network Function Virtualization (NFV) make dynamic automation and manipulation of the network assets through software-based management possible.

This provides intelligent cities with scalability and flexibility required to address emerging needs and enhance network performance to the fullest extent. For [15], these technologies support rapid deployment of emerging network services.

Relevance to Sustainable Smart Cities

The construction of green smart cities is supported by the enabling characteristics of 5G, such as high data rates, low latency, and massive device connections. These attributes facilitate saving waste, optimum utilization of resources, as well as remote monitoring and management of the city's infrastructure. Network slicing provides reliable connectivity for critical infrastructure, URLLC supports autonomous cars, and MTC simplifies the rollout of environmental monitoring sensor networks. 5G technology-based solutions enhancing municipal sustainability are facilitated by 5G and edge computing. The basic properties and features of 5G technology form a revolutionary basis for developing sustainable smart city concepts. 5G enables the deployment of advanced IoT and AI technologies that can address urban sustainability issues by providing faster, more reliable, and more efficient connectivity.

4.3 Transforming Urban Environments Through 5G Technology

Urban 5G deployment is expected to transform the creation of smart, sustainable cities. Due to its super-fast internet rates, minimal latency, and ability to support multiple devices simultaneously, 5G is a fundamental facilitator for a vast array of applications enhancing urban life [16, 17].

This section deals with 5G's influence on urban infrastructure, transportation, health, environmental sustainability, energy management, and public safety (Fig. 4.2).

4.3.1 Enhanced Connectivity and Communication

The capability of 5G to provide previously unheard-of connections is critical to its transformative potential. The successful operation of smart city apps is dependent on real-time data transfer, which 5G can do at rates up to 100 times faster than 4G and with substantially lower latency. This next generation of communications infrastructure enables the proliferation of Internet of Things (IoT) devices, as well as better data collection and processing in cities, resulting in better public services and more informed decision-making.

4.3.2 Smart Infrastructure and IoT Integration

The availability of the 5G network makes IoT devices easily dispersed all over metropolitan infrastructure. The '5G CityBrain' initiative in Granada, Spain, for instance, deploys sensors, 5G connectivity, and advanced artificial intelligence to track tourism, pollution,

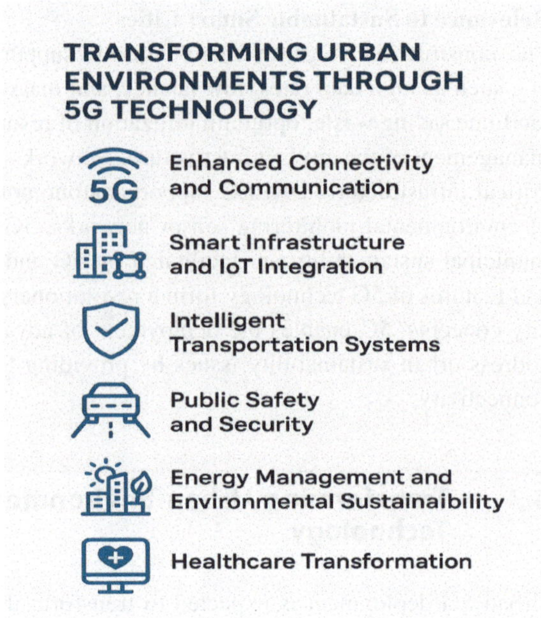

Fig. 4.2 *Source* Author's creation, [18]

and data security to enhance city governance and sustainability. These projects illustrate how 5G facilitates the creation of smart systems that can continuously monitor and maintain vital infrastructure elements like public utilities, bridges, and roads.

4.3.3 Intelligent Transportation Systems

Transport is perhaps the biggest area where 5G can have a vast impact. For autonomous cars, the latency and reliability of smart transportation networks are essential. One of the first glimpses of the future of autonomous public transit are 5G-equipped autonomous buses in cities such as Helsinki and Seoul. 5G enables real-time traffic control through the communication between cars, traffic lights, and sensors, reducing traffic congestion and enhancing road safety.

4.3.4 Public Safety and Security

Public safety projects greatly benefit from improved connection through 5G. Advanced cameras and sophisticated surveillance equipment offer high-definition video in real-time, enabling police officials to monitor public areas more intensively and take prompt action regarding issues. Deploying Internet of Things (IoT)-based environmental monitoring systems that can notify locals of imminent dangers such as air quality issues and natural calamities is easier with 5G and enables the improvement of the city's environment.

4.3.5 Energy Management and Environmental Sustainability

5G technology is critical for better management of energy and promoting environmental sustainability in smart cities. Through enabling real-time interaction between consumers and suppliers of energy, 5G-supported smart grid deployment enhances energy distribution and reduces waste. In addition, 5G-supported smart meters ensure more environmentally friendly behavior by providing residents with in-depth knowledge of how they consume energy. To aid water conservation initiatives, IoT sensor data is utilized to monitor water consumption across the Japanese city of Fukuoka.

4.3.6 Healthcare Transformation

5G integration benefits the healthcare sector tremendously. Ultra-high-speed and zero-latency connectivity make real-time telemedicine possible, facilitating patients to have constant monitoring and remote consultations. In addition, by facilitating the real-time transfer of enormous medical data such as MRI images and X-rays, 5G enhances patient care by accelerating the diagnosis and treatment processes (Fig. 4.3).

4.4 Enabling Speed, Efficiency, and Scalability by Driving 5G Networks

5G deployment seeks to establish a robust, scalable, and efficient digital infrastructure for smart cities alongside improved internet speeds. It supports real-time processing, widespread networking, and effective resource management—all key elements of sustainable urban growth.

4.4.1 Speed: The Real-Time Revolution in Smart Cities

- **Ultra-Low Latency and Data Processing**

5G's very low latency (below 1 ms in the best-case scenario) is ground breaking for real-time use cases such as emergency response systems, traffic monitoring, and AI-based

APPLICATIONS OF 5G IN SMART CITIES AND THEIR BENEFITS

Application Area	Benefits
Enhanced Connectivity	- Ultra-fast internet speeds - Low latency - Support for numerous IoT devices
Smart Infrastructure	- Real-time monitoring of urban assets - Proactive maintenance - Improved public services
Intelligent Transportation	- Support for autonomous vehicles - Reduced congestion
Public Safety	- Advanced surveillance systems - Quick emergency response
Energy Management	- Optimized energy distribution - Reduced waste
Healthcare	- Reliable telemedicine services - Real-time transmission of medical data

Fig. 4.3 Applications of 5G in smart cities and their benefits. *Source* Author's Creation [19]

city operations. It ensures Real-time data relay for public safety and transportation systems, compared to 4G, where network traffic may bog down the transfer of data. Industrial automation and remote healthcare with virtually no delays.

- **Edge Computing for Faster Response Times**

Traditional networks rely on centralized cloud computing, which can introduce lag. 5G integrates edge computing, processing data closer to the source, reducing delays, and ensuring that city-wide IoT operations work smoothly.

4.4.2 Efficiency: Optimizing Urban Resources and Connectivity

Network Slicing for Tailored Services

One of 5G's most innovative features is network slicing, where a single 5G infrastructure can be divided into virtual networks, each serving a specific function. In a smart city, this means:

- Healthcare systems get a secure, high-speed connection for telemedicine.
- Autonomous vehicles receive uninterrupted, low-latency communication.
- Energy grids operate on a stable and highly secure network.

This level of customization ensures efficient bandwidth allocation and resource optimization, unlike the one-size-fits-all approach of previous networks.

Energy Efficiency and Sustainability

5G is built to consume less energy per bit of data transmitted compared to 4G. Key improvements include:

- AI-powered energy management in telecom towers, reducing power consumption.
- Efficient bandwidth allocation, ensuring only required resources are used.
- Longer battery life for IoT devices, reducing electronic waste.

According to a study by the International Telecommunication Union (ITU), 5G networks can be up to 90% more efficient per traffic unit compared to 4G.

4.4.3 Scalability: Building Networks for Growing Cities

Handling Massive IoT Expansion

Smart cities rely on billions of IoT devices for everything from waste management to smart lighting. 5G's enhanced machine-type communication (mMTC) can support up to 1 million devices per square kilometre, ensuring:

- Scalability for growing urban populations without performance drops.
- Seamless integration of new technologies as cities evolve.

Future-Proofing Digital Infrastructure

Unlike its predecessors, 5G is designed to integrate with next-generation technologies, such as:

- 6G and AI-driven urban planning.
- Quantum computing for ultra-secure data transmission.
- Satellite 5G networks for global coverage, even in remote areas.

Private 5G Networks for Enterprises
Many industries within smart cities—such as manufacturing, healthcare, and logistics—are deploying private 5G networks to ensure reliability, security, and scalability. Companies like Siemens and Bosch are already setting up private 5G infrastructures to automate operations.

4.4.4 Governance and Policy Challenges in 5G Deployment

Security and Data Privacy Concerns
With 5G, the sheer volume of connected devices increases cybersecurity risks. Challenges include:

- Higher vulnerability to cyberattacks.
- Data privacy risks with IoT surveillance.
- Potential misuse of AI-driven analytics in governance.

To mitigate these risks, governments need strict data regulations like GDPR in Europe and India's Personal Data Protection Act.

Infrastructure Costs and Digital Divide
The rollout of 5G requires dense network towers and fiber optics, making it expensive and challenging in rural and underdeveloped regions. Governments must:

- Subsidize telecom companies for nationwide 5G coverage.
- Promote public–private partnerships to accelerate deployment.
- Ensure affordable access to 5G services to prevent digital inequality.

Besides being a technology leap, 5G is an economic force for urban development. It lays the ground for smart, robust, and sustainable cities by propelling speed, efficiency, and scalability. Policy structures, cybersecurity, and inclusive infrastructure development must take centre stage in order to make 5G benefits experienced across all sectors of society.

4.5 Concluding Remarks: The Role of 5G in Building Sustainable Smart Cities

The demand for innovative solutions for green development, good public services, and efficient resource allocation is being driven by world urbanization and population. As 5G networks enhance connectivity, support next-generation technologies, enable efficient municipal services, they are vital to the construction of smart cities. The important elements of 5G architecture that support green smart cities include speed, efficiency, scalability, network slicing, virtualization, edge computing, and security capabilities.

4.5.1 5G as the Backbone of Smart Cities

5G technology is the backbone of smart city infrastructure due to its high speed, reliability, and low latency. It facilitates seamless interaction between citizens, devices, and municipal infrastructure, allowing real-time processing and transfer of data to facilitate improved decision-making. Through the automation of public services, smart grids, and smart transportation, the convergence of artificial intelligence (AI) and Internet of Things (IoT) devices enhances smart city solutions [20].

4.5.2 Enhancing Urban Efficiency Through Network Slicing and Virtualization.

Virtualized networks and network slicing enable dedicated networks for an extensive array of civic services. Each business, from energy and transport to public health and safety, has particular requirements that are fulfilled by numerous network slices. Improved services, improved business efficiency, and efficient utilization of municipal resources are all advantages of this approach [21]. For instance, ultra-reliable low-latency communication (URLLC) can benefit emergency services with real-time connectivity during emergencies. Massive Machine Type Communication (mMTC) slices can be utilized to handle large amounts of data in Internet of Things applications like environmental monitoring and traffic management.

4.5.3 Energy Efficiency and Sustainability in Smart Cities

5G's ability to maximize energy consumption is one of its primary advantages in making the urban ecosystem more sustainable and greener. 5G-enabled intelligent grids enable real-time energy supply and monitoring, reducing wastage and improving efficiency. Additionally, with remote energy system management and predictive analysis enabled by 5G networks, renewable energy is made easier to integrate [22].

Apart from power grids, city intelligent lighting systems leverage 5G to dynamically change brightness according to current needs in a bid to conserve energy.

Similar to this, smart water management systems leverage 5G connectivity to identify leaks and optimize water delivery, which addresses issues of the environment.

4.5.4 Accelerating the Adoption of Autonomous Vehicles and Smart Transportation

Widespread adoption of autonomous vehicles and smart transportation systems relies on 5G's extremely low latency and high-speed connectivity. Improved traffic flow, reduced congestion, and increased road safety are enabled through real-time communication among automobiles, traffic infrastructure, and central control systems.

For example, 5G-powered V2X communication allows automotive vehicles to communicate with other vehicles, pedestrian crossings, and traffic lights. By raising fuel efficiency and reducing the likelihood of accidents, this also paves the way for sustainable mobility in cities.

4.5.5 Security and Privacy Considerations in 5G-Powered Smart Cities

With greater and greater use in smart cities, robust security solutions are imperative. With more and more devices, cameras, and critical infrastructure being networked together through Internet of Things connectivity, the threat of hidden vulnerabilities that a hacker could exploit grows. 5G networks employ advanced security solutions such as end-to-end encryption, multi-factor authentication, and artificial intelligence-based threat detection to minimize the risks.

Local planners and governments must enact stringent data privacy regulations to protect residents' data and ensure the ethical use of smart technology.

4.5.6 Challenges and Future Prospects of 5G in Smart Cities

- Despite the much potential that 5G has to create sustainable smart cities, there are a number of hurdles that are hampering its mass adoption:
- ***High-cost infrastructure***: It is very costly to roll out 5G infrastructure like fiber-optic connections and miniature cell networks.

- ***Digital Divide***: Ensuring equal access to 5G connectivity across all socioeconomic groups remains a challenge.
- ***Regulatory and Spectrum Allocation Issues***: Governments must work closely with telecom providers to streamline spectrum allocation and policy frameworks.

5G in smart cities seems to hold a promising future even with all of these barriers. Already, there are proposals to create 6G technology, which will enhance quantum communication, AI-powered automation, and connectivity further. The future smart cities will be influenced, in significant measure, by the continuous evolution of wireless networks, which will make them more effective, secure, and environmentally friendly.

To summarize, 5G is more of an engine of urbanization and not simply a technological progression. 5G accelerates the development of eco-friendly intelligent cities by ensuring faster communication, efficient resource use, and safe operation. Its integration with emerging technologies like AI, blockchain, and IoT will further enhance the urban experience, making cities more livable, efficient, and resilient.

However, to fully harness the potential of 5G, stakeholders—including governments, telecom providers, and technology innovators—must work collaboratively to address infrastructure, regulatory, and security challenges. With strategic implementation and continued advancements, 5G will pave the way for a smarter, more connected, and sustainable future.

References

1. Kachhavay, M. G., & Thakare, A. P. (2014). 5G technology-evolution and revolution. *International Journal of Computer Science and Mobile Computing*, *3*(3), 1080-1087.
2. Mishra, P., & Singh, G. (2023). Sustainable Smart Cities. *Enabling Technologies, Energy Trends and Potential Applications.*
3. Andrews, J. G., Buzzi, S., Choi, W., Hanly, S. V., Lozano, A., Soong, A. C., & Zhang, J. C. (2014). What will 5G be?. IEEE journal on selected areas in communications, 32(6), 1065–1082.
4. Boccardi, F., Heath Jr, R. W., Lozano, A., Marzetta, T. L., & Popovski, P. (2014). Five disruptive technology directions for 5G. IEEE communications magazine, 52(2), 74–80.
5. Shah, P. (2023). The Future of IoT and 5G: Unlocking New Possibilities in Connectivity. Advances in Computer Sciences, 6(1)
6. Liu, C., Shang, J., & Liu, C. (2023). Exploring household food waste reduction for carbon footprint mitigation: a case study in Shanghai, China. Foods, 12(17), 3211.
7. Ratasuk, R., & Mangold, S. (2016). Balancing performance and energy efficiency in 5G wireless networks. IEEE Wireless Communications, 23(4), 88–95.
8. Shafi, M., Liu, X., Frontera, R., García-Navarro, J. J., Perraud, E., Zhang, M., ... & Imran, M. A. (2018). 5G for future vertical industries: Requirements and challenges. IEEE communications magazine, 56(2), 112–119.
9. 3GPP. (2017). Study on scenarios and requirements for next generation access technologies (Technical Report No. TR 38.913). 3rd Generation Partnership Project.

10. 3GPP. (2019). Service requirements for User Equipment (UE) in idle mode (Technical Specification No. TS 22.104). 3rd Generation Partnership Project.
11. 3GPP. (2019). System architecture for the 5G System; Stage 2 (Technical Specification No. TS 23.501). 3rd Generation Partnership Project.
12. 3GPP. (2018). Study on architecture aspects for 5G Public Warning System (PWS) (Technical Report No. TR 23.720). 3rd Generation Partnership Project.
13. Larsson, E. G., Edfors, O., Doré, J., & Johansson, R. (2014). Massive MIMO for next generation wireless systems. IEEE communications magazine, 52(2), 186–195.
14. Checko, A., Christiansen, H. L., Yan, Y., Scolari, L., Kardaras, G., Berger, M. S., & Dittmann, L. (2015). Cloud RAN for mobile networks: a technology overview. IEEE Communications Surveys & Tutorials, 17(1), 405–426.
15. Kreutz, D., Ramos, F. M., Verissimo, P., Rothenberg, C. E., Azodolmolky, S., & Uhlig, S. (2015). Software-defined networking: A comprehensive survey. Proceedings of the IEEE, 103(1), 14–76.
16. Regnauld, T. (2024, November 22). The impact of 5G technology on smart cities: Paving the way for the future. Medium. https://medium.com/%40theophileregnauld941/the-impact-of-5g-technology-on-smart-cities-paving-the-way-for-the-future-0c0868915717
17. Yang, C., et al. (2022). Using 5G in smart cities: A systematic mapping study. *Intelligent Systems with Applications, 14*, 200065.
18. George, A. S. (2024). 5G-Enabled Digital Transformation: Mapping the Landscape of Possibilities and Problems. Partners Universal Innovative Research Publication, *2*(3), 01–37.
19. Rao, S. K., & Prasad, R. (2018). Impact of 5G technologies on smart city implementation. *Wireless Personal Communications*, 100(1), 161–176.
20. International Telecommunication Union (ITU). (2024). 5G and Smart Cities: Enabling Sustainable Urban Development. Retrieved from https://www.itu.int
21. GSMA. (2023). 5G Network Slicing and Virtualization for Smart Cities. Retrieved from https://www.gsma.com
22. Ericsson. (2023). The Role of 5G in Energy Efficiency and Sustainability. Retrieved from https://www.ericsson.com

Case Studies

5

5.1 Case Study 1: Schneider Electric's AI-Driven Smart Metering System

The world leader in energy management and automation, Schneider Electric, has made substantial strides toward the creation of green power grids with its AI-based smart metering program, which is solidly built on its EcoStruxure Grid platform. This path breaking system—now in use in North America, Europe, and Singapore—was designed to combat some of the energy industry's most pressing problems: reducing energy costs for consumers; accommodating ever-larger quantities of renewable energy sources; and keeping the electricity supply grid stable and safe. The whole effort is in line with Schneider Electric's first strategic goal: using best-in-class, cutting-edge technologies to craft reliable, sustainable energy systems.

This system is integrated with technologies of artificial intelligence and machine learning. These convert conventional models of power distribution into data-driven, dynamic, responsive networks. If you want something to be smart, it needs to talk. So the first step in making the electric power system smarter was installing communication-embedded, real-time, smart power meters that could report back to the system. These facilities have been established at a number of utility companies, and several working groups are working in unison across this project to prototype the smart meter, with various features, in an actual system. After installation, the AI project needs time to bet on the basic infrastructure of the electric power grid.

The grid optimization made possible by these predictive signals preserves effective energy use without significantly impacting customer behavior. Schneider's smart demand response systems also modulate energy delivery through dynamic pricing and real time grid equilibrium. Customers are encouraged to shift their usage off-peak, alleviating stress on infrastructure and facilitating more even, balanced grid operation.

V. P. Gupta et al., *Green IoT and AI for Sustainable Development of Smart Cities*,
Synthesis Lectures on Sustainable Development,
https://doi.org/10.1007/978-3-031-98743-4_5

The EcoStruxure Grid is unique in that it is not only centered on the convergence of renewable energy but also on customer engagement. Customers are provided with personalized analysis and graphical tools to keep track of and optimize their energy usage habits via an easy-to-use digital portal on PC and mobile devices. Such features voluntarily encourage energy-saving measures along with awareness.

The software solutions developed by Schneider implement effective solutions for the unpredictable nature of renewable energy sources. Dynamic storage mechanisms operated by the system allow peak production times to be utilized for renewable energy distribution thus avoiding fossil fuel consumption. The reduction of carbon emissions in the power sector through Schneider's efforts stems from a perfect fusion of renewable energy technology with smart metering solutions. Due to smart implementation investors have obtained substantial benefits.

End users achieve up to 12% cost savings in their electricity expenses when they move their power usage to off-peak hours while the peak demand drops by 15% which enables a more reliable power grid operation. The environmental benefits are major because energy wastage reduction results in decreased carbon emissions which help achieve sustainable global targets. The system's scalability and potential to address new energy requirements depends on its dependence on cloud computing together with big data analytics capabilities. The EcoStruxure Grid functions as an extensible next-generation utility system design which depends on predictive maintenance and real-time reactivity as well as real-time monitoring. Schneider Electric continues to invest in advanced technologies for the future development of its smart metering solutions.

Paradigms of energy distribution and consumption are soon to be redefined by technologies like AI-controlled micro grids, peer-to-peer energy trading via block chain, and smart electric vehicle charging coordination. AI will have a more prominent role in managing dynamic load distribution with the rise in the usage of electric vehicles. To avoid grid overload and allow for more frictionless adoption of electric vehicles, smart meters will help to synchronize EV charging in off-peak hours. AI-based autonomous management will aid micro grids, particularly rural or industrial, by ensuring cost-effectiveness and reliability of energy. In short, Schneider Electric's AI-powered smart metering solution is a leap towards smart, efficient, and sustainable energy management. Though issues of data privacy, upfront cost of implementation, and hardware upgrades remain, the proven advantages in saving costs, grid resilience, and carbon savings are too significant to overlook. As the needs for global energy change, the combination of AI with smart metering and grid technologies shall be the foundation of sustainable energy solutions, and both consumers and utility providers become leaders in the energy revolution.

Discussion Questions

1. How has Schneider Electric's AI-driven smart metering system improved energy efficiency and reduced electricity costs for consumers?

2. What are the key benefits of using AI in demand response and smart metering, and how do they contribute to grid stability?
3. What challenges do AI-driven smart metering systems face in terms of implementation, infrastructure, and data security?

5.2 Case Study 2: Electric Vehicles and AI – Driven Sustainable Urban Mobility with Tesla

Tesla has become a global leader in reshaping urban transportation through the infusion of artificial intelligence (AI) in electric mobility, building an ecosystem that transcends the making of electric vehicles (EVs). With sustainability as its core purpose, Tesla uses AI to drive smarter, cleaner cities by making its vehicles and infrastructure safer, more efficient, and more adaptable. By leveraging AI-based technologies like autonomous driving, intelligent energy management, and predictive maintenance, Tesla is not just improving the functionality of electric vehicles but also leading the broader transformation of urban transportation networks.

AI-based system optimization lies as the key driver behind this revolution since it allows Tesla to create integrated platforms for cities together with drivers.

The prominent use cases of AI emerge from Tesla Autopilot along with Full Self-Driving (FSD) attributes. The system constructs an instant high-definition image of surrounding car areas through its configuration of multiple radar devices and camera sensors and GPS modules. System monitoring continuously occurs in Tesla vehicles while the software learns about traffic obstacles and responds to newly emerging situations with human-level precision through deep learning expressions. Each new Tesla generates more accurate neural net performance for the FSD system as it collects driving data which the company uses to make safety improvements and decrease accident probability particularly in cities where random road behavior prevails.

Cars utilize this self-developing software to communicate with traffic signals and execute instant choices that enhance traffic connectivity as well as reduce delays. The combination of AI technologies within Tesla vehicles enhances both safety and efficiency in busy city driving areas by reducing human mistakes while optimizing driver assistance tools.

The application of Artificial Intelligence participates actively in the development of sustainable operations as well as energy-efficient systems. AI in Tesla Cars utilize scanners to detect individual driving patterns along with traffic conditions as well as roadway conditions to maximize operational efficiency optimize battery efficiency.

The intelligent management systems integrate these technologies to minimize their impact on performance. The technology uses AI to enhance battery runtime as well as minimize power consumption thus benefiting the environment. The Tesla smart charging network uses predictive modeling together with load-balancing algorithms for its operations.

Supercharger network, to distribute power. Real-time station capacity evaluation determines the most efficient routing for drivers to wait less and receive maximum energy benefits.

AI predictions about residential energy needs combined with grid statuses enhance Tesla home charging through solar panel and Powerwall integration. This integration leads to a sustainable energy cycle because it minimizes fossil fuels and stimulates renewable power initiatives.

Tesla enables customers to receive software updates automatically through their Over-the-Air system. These car systems develop through time by receiving security feature and performance-based upgrades from software patches. Adjustments to entire new features. AI monitors driving behavior, usage patterns, and Auto manufacturers utilize environmental conditions alongside specific cars to create customized software updates called patches. Traffic patterns allow Tesla cars to receive new city navigation capability while cold-weather conditions benefit from optimized battery performance. Every car optimizes its usefulness and lifecycle duration through the process of continuous adaptation sharing and learning. The system decreases replacement expenses and total environmental footprint because of ownership.

Beyond the immediate vehicle design Tesla holds additional plans for city cars which extend past what you can observe in the vehicle. Tesla sits in a unique.

The company holds a distinctive position to assist urban planners during transportation system development through its data collection and analysis. The company gathers and analyzes immense amounts of real-time data collected from drives. The company collects significant data points about traffic patterns alongside EV charging practices and transit congestion trends.

These metrics help determine enhanced transportation systems through their beneficial application. The city requires enhanced public transit lines which need improvements through upgrades or requires completely new construction investment. The transition to autonomous EV fleets alongside ride-sharing services helps Tesla technology enhance road utilization without affecting service effectiveness for the future city transportation system. AI-powered car data guides policymakers in creating sustainable policies that enhance both environmental quality and city air cleanliness. The energy solution packages from Tesla deal with both the specific challenge of grid reliability and the general issue of power network strength. Tesla advocates decentralized power generation and utilization through the integration of AI, EVs, solar power, and energy storage. Artificial intelligence systems enable the delivery of clean power throughout any area at any moment. The process of weather forecasting along with household energy management and grid demand adjustments is enabled by artificial intelligence systems. The energy paradigm works towards developing an adaptive system which balances renewable energies to push forward renewable transition goals. The increasing number of urban dwellers together with widespread electric vehicle usage leads to increased power requirements for cities. The AI-based technology from Tesla provides a manageable resource management system through which utility needs can scale without needing additional power plants.

Tesla has come a long way, but there is much more to go in its global rollout of AI-driven EV infrastructure. Some of the major challenges that still exist are national law differences, insufficient charging points in certain regions, and privacy issues. Cultural and economic preparedness also come into play, with developing countries facing further challenges in terms of digital access and energy availability. Tesla will need to keep working with governments, utilities, and city stakeholders to create safe and equitable systems that drive mass take-up if it is to overcome these challenges. Bringing equal access to AI-enabled mobility within reach will need international cooperation, infrastructure investment, and public education.

In short, Tesla is creating a new standard for sustainable urban mobility by bringing together AI and electric cars. With its intelligent energy systems, real-time application of data, autonomous driving, and ongoing optimization of vehicles, Tesla shows the world how technology can enhance transport to be safer, cleaner, and more efficient. Tesla's business model gives us a vision for the future of how AI and electrification will revolutionize cities, as other energy and auto companies take notice. With ongoing innovation and a focus on sustainability, AI-powered EV ecosystems can drive the way toward smarter, greener, and more livable cities.

Discussion Questions

1. How does Tesla utilize AI to improve the safety and efficiency of its electric vehicles in urban environments?
2. In what ways can data collected from AI-powered EVs assist urban planners in designing better transport systems?
3. What role does AI play in energy optimization for EV charging and grid management?

The manufacturer's authorised representative in the EU is Springer
Nature Customer Service Centre GmbH, Europaplatz 3, 69115 Heidelberg,
Germany. If you have any concerns regarding our products, please
contact ProductSafety@springernature.com

Printed and bound by CPI Group (UK) Ltd, Croydon, CR0 4YY
30/09/2026
02225577-0018